Mastering Essential Math Skills

20 minutes a day to success

Book Two: Middle Grades/High School
3rd EDITION

Richard W. Fisher

Edited by Christopher Manhoff

For access to the
Online Video Tutorials,
to go <u>www.mathessentials.net</u>
and click on Videos.

★★★★★ **Free Gift for a 5-Star Amazon Review!**

We hope you enjoy this book. We love 5-star Amazon reviews. If you love this book and give it a 5-star review, as a token of our appreciation we would like to present you with a free gift. Just email us at **math.essentials@verizon.net** and let us know the name you used to post the review, and which book you reviewed, and we here at Math Essentials will send you a free gift (a $22.95 value).

Tell your friends about our award-winning math materials, like us on Facebook and follow us on Twitter. You are our best advertising, and we appreciate you!

Math Essentials

Mastering Essential Math Essentials Book 2, 3rd Edition

Manufactured in the United States of America

ISBN: 978-0-9994433-8-5

1st printing 2018

Math Essentials
P.O. Box 1723
Los Gatos, CA 95031
866-444-MATH (6284)
www.mathessentials.net
math.essentials@verizon.net

Notes to the Teacher or Parent

What sets *Mastering Essential Math Skills* apart from other books is its approach. It is not just a math book, but a *system* of teaching math. Each daily lesson contains five key parts: two speed drills, review exercises, teacher tips (Helpful Hints), a section containing new material, and a daily word problem. Teachers have flexibility in introducing new topics, but the book provides them with the necessary structure and guidance. The teacher can rest assured that essential math skills are being systematically learned.

With so many concepts and topics in the math curriculum, some of these essential skills are easily overlooked. This easy-to-follow math program requires only twenty minutes of instruction per day. Each lesson is concise, and self-contained. The daily exercise help students to not only master math skills, but also to maintain and reinforce those skills through consistent review—something that is missing in most math programs. Skills learned in this book apply to all areas of the math curriculum, and consistent review is built into each daily lesson. Teachers and parents will also be pleased to note that the lessons are quite easy to correct.

The book is divided into eight chapters, which cover whole numbers, fractions, decimals, percentages, integers, geometry, charts and graphs, and problem solving.

Mastering Essential Math Skills is based on a system of teaching that was developed by a math instructor over a twenty-year period. This system has produced dramatic results for students. The program quickly motivates students and creates confidence and excitement that leads naturally to success.

Please read the following "How to Use This Book" section and let this program help you to produce dramatic results with your children or math students.

How to Use this Book

Mastering Essential Math Skills is best used on a daily basis. The first lesson should be carefully gone over with the students to introduce them to the program and familiarize them with the format. A typical lesson has been broken down on the following pages into steps to suggest how it can best be taught. It is hoped that the program will help your students to develop an enthusiasm and passion for math that will stay with them throughout their education.

As you go through these lessons every day you will soon begin to see growth in the students' confidence, enthusiasm, and skill level. The students will maintain their mastery through the daily review.

In school, the book is best used during the first part of the math period. The structure and format seem to naturally condition the students to "think nothing but math" from the moment class begins. The students are ready to "jump into the lessons" without any prompting or motivating needed from the teacher. This makes for a very smooth and orderly start each day.

Also, once you have finished the daily lesson, there will still plenty of time to explain related topics, or work on new topics in the basic test or through other sources.

Step 1

Students open their books to the appropriate lesson and begin together. Have students first go to the review exercises, working each problem and showing all their work. If students finish early, they are to check their work in the review section.

Step 2

When you feel that enough time has been spent on the review exercises (usually two to three minutes), call out "Time." The next step is to go to the speed drills. A good signal is to say, "Get ready to add." The students go to the addition drill and wait for the next signal. Then say, "The number to add is '( )'." At this stage the students place the given number inside the addition circle and, as quickly as possible, write all the sums in the appropriate space outside the perimeter of the circle. As students complete the drill, have them drop their pencils and stand or signal in some appropriate way. When enough time has been given, say, "Time." Students then correct the drill as the answers are read aloud by the teacher or a student. The same process is used for the multiplication drill. It is amazing how motivating these speed drills can become in helping students to master their addition and multiplication facts.

Step 3

After the speed drills, work through the review problems with the class. Work the problems on the board or overhead and go through them step-by-step with the students, drawing responses and asking questions as you go. Allow the students to check their own work in this section. This section provides consistent review and reinforcement of topics that the class learns.

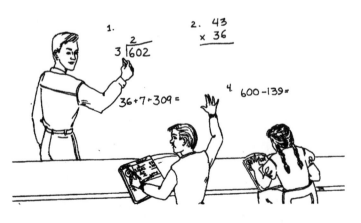

Step 4

After going through the review exercises, give a short introduction of the new material. This is where the teacher's unique style and skills come into play. Appropriate concepts, vocabulary, and skills can be introduced on the blackboard or overhead. This should require only a few minutes.

Step 5

After a brief introduction of the new material, go over the "Helpful Hints" section with the class. Be sure to point out that it is often helpful to come back to this section as the students work independently. This section often has examples that are very helpful to the student.

Step 6

After going through the "Helpful Hints" section, go the two sample problems. It is highly important to work through these two problems with the class. The students can model the techniques that are discussed and demonstrated by the teacher. Go through the steps on the board or overhead, and the students can write them directly into their books. Working these sample problems together with the class can prevent a lot of unnecessary frustration on the part of the students. In essence, in working them together, each student has successfully completed the first two problems of the lesson. This can assist in developing confidence as a routine part of each daily lesson.

Step 7

Next, allow the students to complete the daily exercises and the word problem of the day. Make it a point to circulate and offer individual help. If it is necessary, work another example or two on the board with the entire class. Also, reading the word problem of the day together with the class before they work it independently may be very beneficial.

Step 8

Last, collect the books, correct them and return them the next day. It may sometimes be appropriate to correct them with the students.

Contents

Speed Drills	**Review Exercises**

+

1.   362
   +   75

2.   723
   +   12

x

3.   72
   x   3

4.   6
          4
   +   2

Helpful Hints	1. Line up numbers on the right side. 2. Add the ones first. * "sum" means to total or add 3. Remember to regroup when necessary.

S.   342
        63
  +   512

S.   436
       632
  +   416

1.   42
        53
  +   16

2.   716
        72
  +   314

1	
2	

3.   616
       724
       642
  +   16

4.   723
       436
        19
  +   8

5.   346
       453
       964
  +   234

6.   6
        17
       418
  +   234

3	
4	
5	
6	
7	

7. 362 + 436 + 317 =

8. 27 + 44 + 63 + 73 =

9. Find the sum of 334, 616, and 743.

10. Find the sum of 16, 19, 23, and 47.

8	
9	
10	
Score	

Problem Solving	There are 33 students in Mr. Jones' class, 41 students in Ms. Martinez's class, and 26 students in Mr. Kelly's class. How many students are there altogether?

Speed Drills	Review Exercises

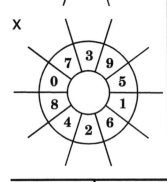

1.
```
      36
      43
   +  16
```

2.
```
     367
      19
   + 437
```

3.   863 + 964 + 17 =

4. Find the sum of 16, 19, 24, and 36

Helpful Hints	When writing large numbers, place commas every three numerals, starting from the right. This makes them easier to read.	**Example:** 21 million, 234 thousand, 416 21,234,416

S.
```
   3,162
     143
  +3,647
```

S.
```
   7,213
     647
  + 22,134
```

1.
```
   1,324,167
 + 2,146,179
```

2.
```
   43,213
    8,137
  +    72
```

3.
```
   75,643
    3,742
  +76,419
```

4.
```
   1,736,412
     136,123
  +    7,423
```

5.
```
    7,163
   14,421
    6,745
 + 14,123
```

6.
```
   63,742
    1,235
   16,347
  + 2,425
```

1	
2	
3	
4	
5	
6	
7	
8	
9	
10	
Score	

7.   Find the sum of 1,342, 176, and 13,417.

8.   17,432 + 7,236 + 6,432 + 16 =

9.   3,672 + 9,876 + 3,712 + 16 =

10.   Find the sum of 72,341, 2,342, and 7,963

Problem Solving	One state has a population of 792,113, and another state has a population of 2,132,415, and a third state has a population of 1,615,500. What is the total population of all three states?

Whole Numbers   Subtraction

Speed Drills	Review Exercises

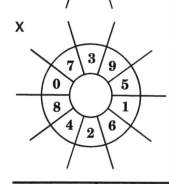

+

1.   3,163
            424
      + 6,734

2. Find the sum of 1,234, 32,164, and 7,321

×

3.   76,723
         15,342
      + 6,412

4. 3,426 + 73,164 + 17 + 17,650 =

Helpful Hints	1. Line up the numbers on the right side. 2. Subtract the ones first. 3. Regroup when necessary.	* "find the difference" means to subtract " is how much more" means to subtract

S.   643
  – 162

S.   3,224
  – 763

1.   312
  – 71

2.   716
  – 317

3.   7,163
  – 778

4.   7,121
  – 1,244

5.   1,493
  – 846

6.   4,492
  – 1299

1	
2	
3	
4	
5	
6	
7	
8	
9	
10	
Score	

7. Find the difference between 124 and 86.

8. Subtract 426 from 3,496.

9. 7,146 – 747 =

10. 986 is how much more than 647?

Problem Solving

712 students attend Monroe School and 467 students attend Jefferson School. How many more students attend Monroe School than Jefferson School?

Speed Drills	Review Exercises

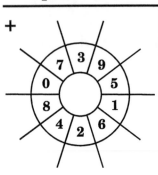

+

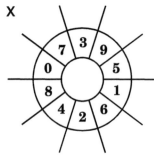

×

1.
```
    376
    427
 +  363
```

2.   716 - 142 =

3. 3,172 + 76 + 7,263 =

4.
```
   6,413
 -   764
```

Helpful Hints	1. Line up numbers on the right. 2. Subtract the ones first. 3. Sometimes a number must be regrouped more than once.	**Examples:**

Examples:
```
 6 10 9 1
 7,1 0 3
 -  6 7 7
 6,4 2 6
```
```
 5 9 9
 6,0 0 0
 -1,6 3 4
 4,3 6 6
```

S.
```
    701
 -  267
```

S.
```
    600
 -  379
```

1.
```
     70
 -   54
```

2.
```
    603
 -  168
```

3.
```
   5,013
 - 1,405
```

4.
```
    500
 -  236
```

5.
```
   7,000
 -   634
```

6.
```
   3,102
 - 1,634
```

7. Subtract 7,632 from 12,864.

8. Find the difference between 13,601 and 76,021.

9. 76,102 − 63,234 =

10. What number is 7,001 less than 9,000?

1	
2	
3	
4	
5	
6	
7	
8	
9	
10	
Score	

Problem Solving	A company earned 96,012 dollars in its first year and 123,056 dollars in its second year. How much more did the company earn in its second year than in its first year?

8

Speed Drills	Review Exercises

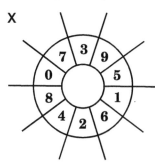

+

x

1.   701
     − 637

2.   7,102
     − 673

3.   343,672
        72,164
   + 736,243

4. $5,000 - 467 =$

Helpful Hints   Use what you have learned to solve the following problems.

1	
2	
3	
4	
5	
6	
7	
8	
9	
10	
Score	

S.   743
     7,614
    16,321
   + 5,032

S.   5,001
   − 1,346

1.   346
        25
    + 176

2.   716
     − 143

3.   4,216
       764
    + 5,123

4. 3,732,246
  + 3,510,762

5.   7,101
    − 1,436

6. Find the differ-
    ence between
    1,964 and 768.

7.   17,023 − 13,605.

8.   Find the sum of 236, 742, and 867.

9.   How much more is 763 than 147?

10.   72,163 + 16,432 + 1,963 =

Problem Solving   A theater has 905 seats. If 693 of them are taken, how many seats are empty?

Speed Drills	**Review Exercises**

+

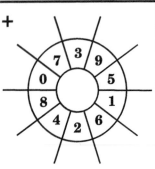

1.   6,032
     − 1,647

2. 364 + 72 + 167 + 396 =

x

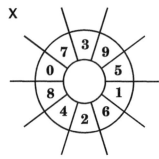

3. Find the difference between 760 and 188.

4.   9,000 - 3,287 =

Helpful Hints	1. Line up numbers on the right. 2. Multiply the ones first. 3. Regroup when necessary. 4. "Product" means to multiply.	**Examples:**	$\overset{1\ 1}{644}$ $\underline{\text{x}\qquad 3}$ $1,932$	$\overset{3\ 2}{6,076}$ $\underline{\text{x}\qquad 4}$ $24,304$

S.   423   S.   2,345   1.   67   2.   74
    x   3   x   6   x   3   x   6

3.   764   4.   3,142   5.   2,036   6.   3,427
    x   3   x   6   x   8   x   6

7.   8,058 x 7 =

8.   7 x 7,643 =

9.   Find the product of 9,746 and 6.

10. Multiply 9 and 13,708.

1	
2	
3	
4	
5	
6	
7	
8	
9	
10	
Score	

Problem Solving	If there are 365 days in each year, then how many days are there in six years?

Speed Drills	Review Exercises

+

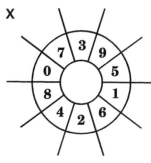

x

**Helpful
Hints**

1.   342
   x   6

2.   2034
    x   7

3.   7,103
    − 1,664

4.   32,173
         1,424
     + 3,456

1. Line up numbers on the right.   **Examples:**   43   437
2. Multiply the ones first.   x 32   x 26
3. Multiply by tens second.   86   2622
4. Add the two products.   1290   8740
                                                     1,376   11,362

S.   46   S.   146   1.   73   2.   75
   x 23   x 42   x 62   x 16

3.   47   4.   534   5.   237   6.   807
   x 36   x 25   x 30   x 36

1	
2	
3	
4	
5	
6	
7	
8	
9	
10	
Score	

7.   Find the product of 16 and 28

8.   38 x 26 =

9.   Find the product of 92 and 734

10.   320 x 49 =

Problem Solving	A school has twenty-six classrooms. If each classroom needs 32 desks, how many desks are needed altogether?

Speed Drills

+

x

Helpful Hints

Review Exercises

1.
$$\begin{array}{r} 23 \\ \times\ 46 \\ \hline \end{array}$$

2.
$$\begin{array}{r} 402 \\ \times\ 36 \\ \hline \end{array}$$

3.
$$\begin{array}{r} 7,205 \\ -\ 1,637 \\ \hline \end{array}$$

4.
$$\begin{array}{r} 335 \\ 63 \\ 426 \\ +\ 173 \\ \hline \end{array}$$

1. Line up numbers on the right.
2. Multiply the ones first.
3. Multiply by tens second.
4. Multiply by hundreds last.
5. Add the products.

Examples:

$$\begin{array}{r} 243 \\ \times\ 346 \\ \hline 1458 \\ 9720 \\ 72900 \\ \hline 84,078 \end{array}$$

$$\begin{array}{r} 673 \\ \times\ 307 \\ \hline 4711 \\ 0000 \\ 201900 \\ \hline 206,611 \end{array}$$

S.
$$\begin{array}{r} 132 \\ \times\ 234 \\ \hline \end{array}$$

S.
$$\begin{array}{r} 623 \\ \times\ 542 \\ \hline \end{array}$$

1.
$$\begin{array}{r} 233 \\ \times\ 215 \\ \hline \end{array}$$

2.
$$\begin{array}{r} 326 \\ \times\ 514 \\ \hline \end{array}$$

3.
$$\begin{array}{r} 143 \\ \times\ 203 \\ \hline \end{array}$$

4.
$$\begin{array}{r} 246 \\ \times\ 403 \\ \hline \end{array}$$

5.
$$\begin{array}{r} 324 \\ \times\ 616 \\ \hline \end{array}$$

6.
$$\begin{array}{r} 263 \\ \times\ 300 \\ \hline \end{array}$$

7. $361 \times 423 =$

8. Find the product of 306 and 427

9. Multiply 600 and 721

10. $334 \times 466 =$

1	
2	
3	
4	
5	
6	
7	
8	
9	
10	
Score	

Problem Solving

A factory can produce 215 cars per day. How many cars can it produce in 164 days?

Whole Numbers   Reviewing Multiplication of Whole Numbers

Speed Drills	Review Exercises

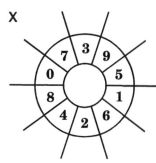

+

x

1.   306
     x 7

2. Find the product of 24
    and 36

3.   7,736
        493
      +2,615

4. Find the difference
    between 2,174 and 636

Helpful Hints	Use what you have learned to solve the following problems.

S.   316
      x 24

S.   604
     x 423

1.   27
       x 6

2.   603
       x 7

3.   3,612
       x 9

4.   63
      x72

5.   263
      x 54

6.   242
     x 643

7. Find the product of 12 and 473

8. 600 x 748 =

9. 22 x 410 =

10. 706 x 304 =

1	
2	
3	
4	
5	
6	
7	
8	
9	
10	
Score	

Problem Solving	Each package of paper contains 500 sheets. How many sheets of paper are there in 24 packages?

Speed Drills	Review Exercises

+

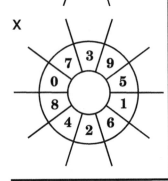

x

Helpful Hints

1. Divide.
2. Multiply.
3. Subtract.
4. Begin again.

Examples:

$$3\overline{)47} \; {}^{15\,r2}$$
$$-\,3\!\downarrow$$
$$\overline{17}$$
$$-\,15$$
$$\overline{2}$$

$$6\overline{)59} \; {}^{9\,r5}$$
$$-\,54$$
$$\overline{5}$$

REMEMBER! The remainder must be less than the divisor.

Review Exercises

1. 712
    $-\,463$

2. 320
    $\underline{x\,6}$

3. $65{,}426$
    $+\,73{,}437$

4. Find the product of 26 and 37

S. $3\overline{)16}$ S. $7\overline{)69}$ 1. $4\overline{)34}$ 2. $8\overline{)43}$

3. $7\overline{)87}$ 4. $6\overline{)93}$ 5. $8\overline{)97}$ 6. $6\overline{)43}$

7. $66 \div 7 =$

8. $97 \div 4 =$

9. $\dfrac{61}{5} =$

10. $\dfrac{37}{2} =$

1	
2	
3	
4	
5	
6	
7	
8	
9	
10	
Score	

Problem Solving

A teacher needs 72 rulers for his class. If rulers come in boxes that contain six rulers, how many boxes of rulers does the teacher need?

Whole Numbers   Dividing Hundreds by 1-Digit Divisors

Speed Drills	Review Exercises

+

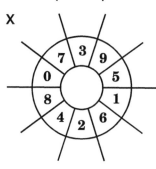

X

1. $6\overline{)39}$ 2. $5\overline{)79}$

3. $7 \times 2{,}134 =$ 4. $6{,}000 - 768 =$

Helpful Hints

1. Divide.
2. Multiply.
3. Subtract.
4. Begin again.

Examples:

$$
\begin{array}{r}
171\,r2 \\
3\overline{)515} \\
-3\downarrow\downarrow \\
\hline
21 \\
-21 \\
\hline
05 \\
-3 \\
\hline
2
\end{array}
$$

$$
\begin{array}{r}
203 \\
4\overline{)812} \\
-8\downarrow\downarrow \\
\hline
01 \\
-0 \\
\hline
12 \\
-12 \\
\hline
0
\end{array}
$$

REMEMBER! The remainder must be less than the divisor.

S. $3\overline{)432}$ S. $6\overline{)237}$ 1. $3\overline{)952}$ 2. $2\overline{)819}$

3. $7\overline{)924}$ 4. $6\overline{)817}$ 5. $5\overline{)727}$ 6. $6\overline{)950}$

7. $4\overline{)484}$ 8. $9\overline{)886}$ 9. $8\overline{)979}$ 10. $6\overline{)953}$

1	
2	
3	
4	
5	
6	
7	
8	
9	
10	
Score	

Problem Solving

A theater has 325 seats They are placed in 9 equal rows. How many seats are in each row? How many seats will be left over?

Speed Drills	**Review Exercises**

+

x

1. $3\overline{)79}$ 2. $6\overline{)557}$

3.   46   4. Find the difference
      x 23   between 236 and 84

Helpful Hints

Examples:

1. Divide.
2. Multiply.
3. Subtract.
4. Begin again.

```
       1708
    3 | 5124
    - 3↓↓↓
       21
     - 21
       02
      - 0
       24
     - 24
        0
```

```
      448 r2
    4 | 1794
    - 16↓↓
       19
     - 16
       34
     - 32
        2
```

REMEMBER! The remainder must be less than the divisor.

S. $3\overline{)7062}$ S. $4\overline{)3452}$ 1. $2\overline{)1132}$ 2. $2\overline{)6743}$

3. $7\overline{)7854}$ 4. $6\overline{)2319}$ 5. $5\overline{)6555}$ 6. $4\overline{)5995}$

7. $7\overline{)73,172}$ 8. $2\overline{)23,960}$ 9. $7\overline{)71,345}$ 10. $6\overline{)32,106}$

1	
2	
3	
4	
5	
6	
7	
8	
9	
10	
Score	

Problem Solving

Mrs. Arnold made 1,296 cookies. If she puts them into boxes that contain 9 cookies each, how many boxes will she need?

Speed Drills	**Review Exercises**

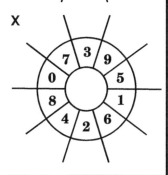

+

x

1. $7\overline{)1422}$

2. $\begin{array}{r} 206 \\ \times\ 36 \\ \hline \end{array}$

3. $\begin{array}{r} 710 \\ -\ 167 \\ \hline \end{array}$

4. 3,752 + 17,343 + 964 =

Helpful Hints	1. Divide. 2. Multiply. 3. Subtract. 4. Begin again.	* Remainders must be less than the divisor * Zeroes may sometimes appear in the quotient

S. $3\overline{)245}$ S. $8\overline{)8568}$ 1. $6\overline{)302}$ 2. $5\overline{)7500}$

3. $3\overline{)6314}$ 4. $9\overline{)3767}$ 5. $7\overline{)1563}$ 6. $8\overline{)716}$

7. $6\overline{)1817}$ 8. $6\overline{)4793}$ 9. $6\overline{)6007}$ 10. $8\overline{)3207}$

1	
2	
3	
4	
5	
6	
7	
8	
9	
10	
Score	

Problem Solving	Six people in a club each sold the same number of tickets. If 636 tickets were sold, how many tickets did each person sell?

Speed Drills	Review Exercises

+

x

Helpful Hints

1. 7⟌818

2. 2,003 − 765 =

3.   453
      x 600

4. 4⟌4003

1. Divide.   **Examples:**
2. Multiply.
3. Subtract.
4. Begin again.

```
           12 r49              44 r5
    60 ⟌ 769            40 ⟌ 1765
       - 60↓               - 160↓
         169                 165
       - 120               - 160
          49                   5
```

S. 30⟌187   S. 40⟌5342   1. 70⟌342   2. 60⟌399

3. 50⟌463   4. 50⟌727   5. 30⟌1763   6. 20⟌1423

7. 50⟌8324   8. 90⟌9281   9. 50⟌4751   10. 50⟌2526

1	
2	
3	
4	
5	
6	
7	
8	
9	
10	
Score	

Problem Solving

A bakery put cookies into boxes of 20 each. If 1,720 cookies were baked, how many boxes would be needed?

Speed Drills	**Review Exercises**

+

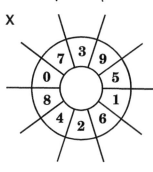

x

Helpful Hints

1. 643
    76
    + 492

2.   403
    − 247

3. 675
    x 32

4. 7 ⟌ 8172

Sometimes it is easier to mentally round the divisor to the nearest multiple of ten.

Example:

$$32\,r11$$
$$22\,⟌\,715$$
$$-\,66↓$$
$$55$$
$$-\,44$$
$$11$$

Think of

$$20\,⟌\,715$$

S. 31 ⟌ 672   S. 22 ⟌ 684   1. 41 ⟌ 927   2. 28 ⟌ 603

3. 68 ⟌ 743   4. 93 ⟌ 692   5. 43 ⟌ 913   6. 31 ⟌ 984

7. 62 ⟌ 724   8. 43 ⟌ 501   9. 21 ⟌ 946   10. 41 ⟌ 866

1	
2	
3	
4	
5	
6	
7	
8	
9	
10	
Score	

Problem Solving

There are 416 students in a school. If there are 32 students in each class, then how many classes are there?

Speed Drills	**Review Exercises**

+

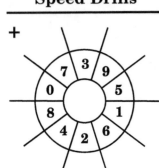

x

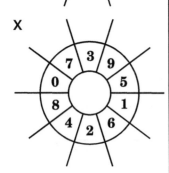

1. $2\overline{)716}$ 2. $30\overline{)524}$

3. $31\overline{)671}$ 4. $39\overline{)791}$

Helpful Hints	Sometimes it is necessary to correct your estimate.

Example:

$$63\overline{)374} \quad \begin{array}{r} 6 \\ \hline \end{array}$$
$$- 378 \text{ -- Too Large}$$

$$63\overline{)375} \quad \begin{array}{r} 5\ r60 \\ \hline \end{array}$$
$$\begin{array}{r} - 315 \\ \hline 60 \end{array}$$

S. $74\overline{)293}$ S. $43\overline{)821}$ 1. $38\overline{)197}$ 2. $87\overline{)522}$

3. $18\overline{)997}$ 4. $21\overline{)178}$ 5. $32\overline{)163}$ 6. $14\overline{)886}$

7. $34\overline{)649}$ 8. $42\overline{)829}$ 9. $36\overline{)721}$ 10. $18\overline{)787}$

1	
2	
3	
4	
5	
6	
7	
8	
9	
10	
Score	

Problem Solving	If each carton contains 36 eggs, how many eggs are there in twenty-four cartons?

Whole Numbers

Dividing Larger Numbers with 2-Digit Divisors

Speed Drills	Review Exercises

+

X

Helpful Hints

1. $30\overline{)96}$

2. $40\overline{)542}$

3. $40\overline{)396}$

4. $57\overline{)361}$

Use what you have learned to solve the following problems. Remember to mentally round your divisor to the nearest multiple of ten.

Example:

$$\begin{array}{r} 216\ r20 \\ 32\overline{)6932} \\ -\ 64\downarrow\downarrow \\ \hline 53 \\ -\ 32 \\ \hline 212 \\ -\ 192 \\ \hline 20 \end{array}$$

Think of

$30\overline{)6932}$

S. $44\overline{)9350}$ S. $31\overline{)6432}$ 1. $23\overline{)2645}$ 2. $41\overline{)9597}$	1
	2
	3
	4
3. $48\overline{)4896}$ 4. $31\overline{)4133}$ 5. $27\overline{)8191}$ 6. $18\overline{)5508}$	5
	6
	7
7. $21\overline{)1537}$ 8. $32\overline{)1286}$ 9. $49\overline{)2222}$ 10. $25\overline{)1049}$	8
	9
	10
	Score

Problem Solving

Sixteen bales of hay weigh a total of 2,000 pounds. How much does each bale weigh if they are all the same size?

Speed Drills	**Review Exercises**

+

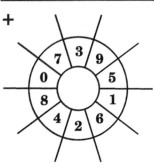

1. $3\overline{)72}$　　　2. $6\overline{)1792}$

X

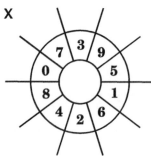

3. $40\overline{)5436}$　　　4. $21\overline{)899}$

Helpful Hints	Use what you have learned to solve the following problems.	**Example:**	* Mentally round 2-digit divisors to the nearest multiple of ten. * Remainders must be less than the divisor.

S. $81\overline{)7816}$　S. $28\overline{)1661}$　1. $2\overline{)87}$　2. $7\overline{)1936}$

3. $5\overline{)5127}$　4. $30\overline{)547}$　5. $70\overline{)196}$　6. $40\overline{)3379}$

7. $38\overline{)9699}$　8. $23\overline{)2633}$　9. $61\overline{)2,222}$　10. $32\overline{)7049}$

1	
2	
3	
4	
5	
6	
7	
8	
9	
10	
Score	

Problem Solving	A car can travel 32 miles using one gallon of gasoline. How many gallons of gasoline will be needed to travel 512 miles?

Reviewing All Whole Number Operations   Name

1.
$$\begin{array}{r} 342 \\ 53 \\ + 616 \\ \hline \end{array}$$

2.
$$\begin{array}{r} 746 \\ 716 \\ 823 \\ + 634 \\ \hline \end{array}$$

3. $7,362 + 775 + 72,516 =$

4. $7,013 + 2,615 + 776 + 29 =$

5. $7,001 + 696 + 18 + 732 =$

6.
$$\begin{array}{r} 743 \\ - 367 \\ \hline \end{array}$$

7.
$$\begin{array}{r} 5,282 \\ - 1,367 \\ \hline \end{array}$$

8. $7,052 - 2,637 =$

9. $6,000 - 3,678 =$

10. $7,001 - 678 =$

11.
$$\begin{array}{r} 76 \\ \times 3 \\ \hline \end{array}$$

12.
$$\begin{array}{r} 7,653 \\ \times 4 \\ \hline \end{array}$$

13.
$$\begin{array}{r} 53 \\ \times 46 \\ \hline \end{array}$$

14.
$$\begin{array}{r} 627 \\ \times 36 \\ \hline \end{array}$$

15.
$$\begin{array}{r} 673 \\ \times 346 \\ \hline \end{array}$$

16. $3\overline{)425}$

17. $6\overline{)1697}$

18. $30\overline{)769}$

19. $42\overline{)8992}$

20. $28\overline{)1577}$

1	
2	
3	
4	
5	
6	
7	
8	
9	
10	
11	
12	
13	
14	
15	
16	
17	
18	
19	
20	

Fractions Identifying

Speed Drills	**Review Exercises**

+

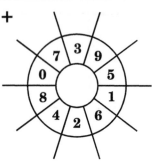

X

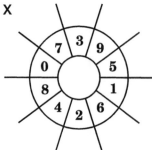

1.
$$701 - 267$$

2.
$$337$$
$$756$$
$$+ \ 63$$

3. $3\overline{)4162}$

4.
$$48$$
$$x \ 27$$

Helpful Hints	A fraction is a number that names a part of a whole or a group	**Example:** $= \dfrac{3}{4} \begin{matrix}\leftarrow\text{numerator}\\ \leftarrow\text{denominator}\end{matrix}$ *Think of $\dfrac{3}{4}$ as $\dfrac{3 \text{ of}}{4 \text{ equal parts}}$

Write a fraction for each shaded figure (some may have more than one name).

1	
2	
3	
4	
5	
6	
7	
8	
9	
10	
Score	

S.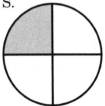

S.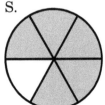

1.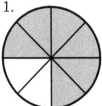

2.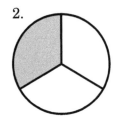

3.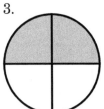

4.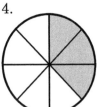

5.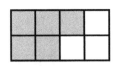

6.

7.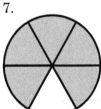

8.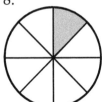

9.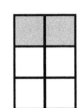

10.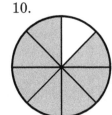

Problem Solving	If a box of crayons holds 24 crayons, how many crayons are there in sixteen boxes?

Speed Drills	Review Exercises

+

x

1. $7 \times 306 =$ 2. $72 + 316 + 726 =$

3. $810 - 316 =$ 4. $20\overline{)317}$

Helpful Hints

$= \dfrac{2}{4} = \dfrac{1}{2}$

$\frac{2}{4}$ has been reduced to its simplest form which is $\frac{1}{2}$. Divide the numerator and denominator by the largest possible number.

Examples: $2\overline{)\frac{6}{8}} = \frac{3}{4}$

Sometimes more than one step can be used:

$2\overline{)\frac{24}{28}} = 2\overline{)\frac{12}{14}} = \frac{6}{7}$

Reduce each fraction to its lowest terms.

S. $\dfrac{5}{10} =$ S. $\dfrac{12}{16} =$ 1. $\dfrac{12}{15} =$ 2. $\dfrac{15}{20} =$

3. $\dfrac{10}{20} =$ 4. $\dfrac{20}{25} =$ 5. $\dfrac{12}{18} =$ 6. $\dfrac{16}{24} =$

7. $\dfrac{24}{40} =$ 8. $\dfrac{20}{32} =$ 9. $\dfrac{15}{18} =$ 10. $\dfrac{18}{24} =$

1	
2	
3	
4	
5	
6	
7	
8	
9	
10	
Score	

Problem Solving

If there are 24 crayons in each box, how many crayons are there in $2\frac{1}{2}$ boxes?

Fractions

Speed Drills	Review Exercises

+

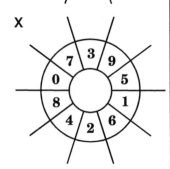

X

1. What fraction of the figure is shaded?

2. Reduce $\frac{6}{9}$ to its lowest terms.

3. $32 \overline{)679}$

4.
$$\begin{array}{r} 216 \\ \times\ 304 \\ \hline \end{array}$$

Helpful Hints

An improper fraction has a numerator that is greater than or equal to its denominator. An improper fraction can be written either as a whole number or as a mixed numeral (a whole number and a fraction).

Example: ⊖⊖⊖◠ $= \frac{7}{2} = 3\frac{1}{2}$

*Divide the numerator by the denominator

$2\overline{)7}$ $\begin{array}{r} 3 \\ 2\overline{)7} \\ 6 \\ \hline 1 \end{array} = 3\frac{1}{2}$

Change each fraction to a mixed numeral or whole number.

S. $\frac{7}{4} =$ S. $\frac{9}{6} =$ 1. $\frac{10}{4} =$ 2. $\frac{10}{7} =$

3. $\frac{30}{15} =$ 4. $\frac{24}{5} =$ 5. $\frac{18}{4} =$ 6. $\frac{36}{12} =$

7. $\frac{45}{10} =$ 8. $\frac{38}{6} =$ 9. $\frac{12}{8} =$ 10. $\frac{60}{25} =$

1	
2	
3	
4	
5	
6	
7	
8	
9	
10	
Score	

Problem Solving

On Saturday, 20,136 people visited the zoo. On Sunday, there were 17,308 visitors. How many more people visited the zoo on Saturday than on Sunday.

Fractions

Speed Drills	Review Exercises

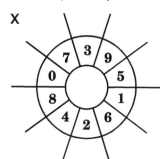

+

X

1. Change $\frac{9}{7}$ to a mixed numeral

2. Change $\frac{25}{20}$ to a mixed numeral

3. Reduce $\frac{25}{35}$ to its lowest terms

4. What fraction of the figure is shaded?

Helpful Hints

To add fractions with like denominators, add the numerators and then ask the following questions: 1. Is the fraction improper? If it is, make it a mixed numeral or whole number. 2. Can the fraction be reduced? If it can, reduce it to its simplest form.

Example:
$$\frac{7}{10}$$
$$+\frac{5}{10}$$
$$\frac{12}{10} = 1\frac{2}{10} = 1\frac{1}{5}$$

S. $\frac{1}{10}$
$+\frac{5}{10}$
_____

S. $\frac{3}{7}$
$+\frac{5}{7}$
_____

1. $\frac{7}{12}$
$+\frac{2}{12}$
_____

2. $\frac{3}{8} + \frac{3}{8}$

3. $\frac{2}{5}$
$+\frac{1}{5}$
_____

4. $\frac{5}{12}$
$+\frac{1}{12}$
_____

5. $\frac{5}{8}$
$+\frac{5}{8}$
_____

6. $\frac{11}{16}$
$+\frac{13}{16}$
_____

7. $\frac{1}{3}$
$\frac{2}{3}$
$+\frac{2}{3}$
_____

8. $\frac{1}{4}$
$\frac{3}{4}$
$+\frac{3}{4}$
_____

9. $\frac{7}{8}$
$+\frac{7}{8}$
_____

10. $\frac{5}{6}$
$+\frac{5}{6}$
_____

1	
2	
3	
4	
5	
6	
7	
8	
9	
10	
Score	

Problem Solving

At Lincoln school $\frac{1}{9}$ of the students ride their bikes to school, $\frac{5}{9}$ of the students walk or ride their bikes to school. What faction of the students either walk or ride their bikes to school? (Be sure your answer is expressed in its simplest form.)

Fractions — Adding Mixed Numerals with Like Denominators

Speed Drills

+

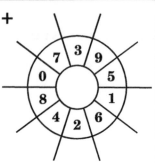

x

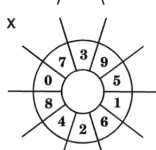

Review Exercises

1.
$$\frac{2}{5}$$
$$+\frac{1}{5}$$

2. $60\overline{)729}$

3.
$$\frac{7}{10}$$
$$+\frac{5}{10}$$

4. $7 + 19 + 342 =$

Helpful Hints

1. Add the fractions first.
2. Add the whole numbers next.
3. It there is an improper fraction, change it to a mixed numeral.
4. Add the mixed numeral to the whole number.

Example: $6\frac{7}{8}$
$+2\frac{5}{8}$

*Reduce fractions to lowest terms

$8\frac{12}{8} = 8 + 1\frac{4}{8} = 9\frac{4}{8} = 9\frac{1}{2}$

S. $3\frac{1}{4}$
$+2\frac{1}{4}$

S. $3\frac{5}{8}$
$+2\frac{3}{8}$

1. $3\frac{2}{5}$
$+4\frac{3}{5}$

2. $4\frac{3}{4}$
$+2\frac{3}{4}$

3. $3\frac{7}{10}$
$+4\frac{3}{10}$

4. $5\frac{5}{6}$
$+2\frac{2}{6}$

5. $6\frac{7}{9}$
$+5\frac{5}{9}$

6. $3\frac{4}{5}$
$+2\frac{2}{5}$

7. $5\frac{9}{10}$
$+3\frac{3}{10}$

8. $7\frac{5}{6}$
$+4\frac{4}{6}$

9. $7\frac{1}{2}$
$+2\frac{1}{2}$

10. $3\frac{5}{7}$
$+4\frac{3}{7}$

1	
2	
3	
4	
5	
6	
7	
8	
9	
10	
Score	

Problem Solving

A baker uses $\frac{7}{8}$ cups of flour for each pie he bakes. He uses $\frac{3}{8}$ cups of flour for each cake. How much flour is used to make 2 pies and 1 cake?

Fractions

Speed Drills	Review Exercises

+

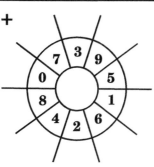

X

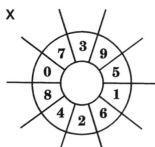

1. Reduce $\frac{15}{18}$ to its lowest terms

2. Change $\frac{19}{5}$ to a mixed numeral

3.
$$\frac{3}{4}$$
$$+\frac{3}{4}$$

4.
$$2\frac{3}{5}$$
$$+\,4\frac{3}{5}$$

Helpful Hints

To subtract fractions that have like denominators, subtract the numerators. Reduce the fractions to their lowest terms.

Example:
$$\frac{9}{10}$$
$$-\frac{3}{10}$$
$$\frac{6}{10} = \frac{3}{5}$$

S.
$$\frac{7}{8}$$
$$-\frac{3}{8}$$

S.
$$\frac{3}{4}$$
$$-\frac{1}{4}$$

1.
$$\frac{5}{8}$$
$$-\frac{1}{8}$$

2.
$$\frac{9}{10}$$
$$-\frac{1}{10}$$

3.
$$\frac{7}{8}$$
$$-\frac{1}{8}$$

4. $\frac{8}{9} - \frac{2}{9} =$

5.
$$\frac{19}{20}$$
$$-\frac{4}{20}$$

6.
$$\frac{7}{11}$$
$$-\frac{3}{11}$$

7.
$$\frac{11}{12}$$
$$-\frac{5}{12}$$

8.
$$\frac{6}{7}$$
$$-\frac{1}{7}$$

9.
$$\frac{23}{24}$$
$$-\frac{13}{24}$$

10.
$$\frac{11}{16}$$
$$-\frac{7}{16}$$

1	
2	
3	
4	
5	
6	
7	
8	
9	
10	
Score	

Problem Solving

John walks $\frac{9}{10}$ miles to school. If he has already gone $\frac{3}{10}$ miles, how much farther does he have to walk before he gets to school?

Speed Drills	Review Exercises

+

X

1.
$$\frac{3}{4}$$
$$-\frac{1}{4}$$

2.
$$\frac{3}{5}$$
$$+\frac{3}{5}$$

3.
$$2\frac{3}{8}$$
$$+3\frac{1}{8}$$

4. $1{,}708 - 765 =$

Helpful Hints

To subtract numerals with like denominators, subtract the fractions first, then the whole numbers. Reduce fractions to lowest terms.

If the fractions can't be subtracted, take one from the whole number, increase the fraction, then subtract.

Examples:

$$7\frac{3}{4}$$
$$-2\frac{1}{4}$$
$$5\frac{2}{4} = 5\frac{1}{2}$$

$$6\frac{1}{4} + \frac{4}{4} = \frac{5}{4}$$
$$-2\frac{3}{4}$$
$$4\frac{2}{4} = 4\frac{1}{2}$$

S. $3\frac{3}{4}$ $\quad -1\frac{1}{4}$

S. $5\frac{1}{3}$ $\quad -2\frac{2}{3}$

1. $7\frac{5}{8}$ $\quad -1\frac{3}{8}$

2. $6\frac{1}{4}$ $\quad -1\frac{3}{4}$

3. $7\frac{7}{15}$ $\quad -2\frac{11}{15}$

4. $8\frac{8}{9}$ $\quad -2\frac{2}{9}$

5. $6\frac{4}{5}$ $\quad -2\frac{2}{5}$

6. $7\frac{1}{10}$ $\quad -3\frac{7}{10}$

7. $4\frac{1}{7}$ $\quad -2\frac{1}{7}$

8. $7\frac{3}{10}$ $\quad -4\frac{7}{10}$

9. $6\frac{1}{5}$ $\quad -2\frac{3}{5}$

10. $7\frac{7}{15}$ $\quad -4\frac{13}{15}$

1	
2	
3	
4	
5	
6	
7	
8	
9	
10	
Score	

Problem Solving

A woman worked $2\frac{2}{3}$ hours on Monday and $3\frac{2}{3}$ hours on Tuesday. How many hours did she work altogether?

| **Speed Drills** | **Review Exercises** |

+

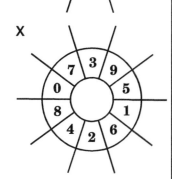

X

1. $\dfrac{1}{4}$

   $+\ \dfrac{1}{4}$

2. $\dfrac{3}{8}$

   $+\ \dfrac{7}{8}$

3. $4\ \dfrac{2}{3}$

   $+\ 3\ \dfrac{2}{3}$

4. $6\ \dfrac{3}{4}$

   $+\ 5\ \dfrac{3}{4}$

| **Helpful Hints** | To subtract a fraction or mixed numeral from a whole number, take one from the whole number and make it a fraction, then subtract. |

Examples:

$\overset{3}{\cancel{4}} \rightarrow \dfrac{4}{4}$
$-\ 2 \quad \dfrac{1}{4}$
$\overline{\quad 1\ \dfrac{3}{4}}$

$\overset{6}{\cancel{7}} \rightarrow \dfrac{5}{5}$
$-\quad \dfrac{3}{5}$
$\overline{\quad 6\ \dfrac{2}{5}}$

S.   6

 $-\ 2\ \dfrac{3}{5}$

S.   7

 $-\quad \dfrac{3}{4}$

1.   6

 $-\ 2\ \dfrac{4}{7}$

2.   5

 $-\ 1\ \dfrac{3}{5}$

3.   7

 $-\quad \dfrac{2}{3}$

4.   6

 $-\ 2\ \dfrac{9}{10}$

5.   7

 $-\ 2\ \dfrac{1}{8}$

6.   16

 $-\ 12\ \dfrac{3}{8}$

7.   7

 $-\ 3\ \dfrac{7}{9}$

8.   4

 $-\ 3\ \dfrac{1}{2}$

9.   6

 $-\ 2\ \dfrac{3}{10}$

10.   5

 $-\quad \dfrac{3}{5}$

1	
2	
3	
4	
5	
6	
7	
8	
9	
10	
Score	

| **Problem Solving** | A tailor had 7 yards of cloth. He used $4\frac{7}{8}$ yards to make a suit. How many yards were left? |

Fractions   Reviewing Adding and Subtracting Mixed Numerals/Like Denominators

Speed Drills

+

X

Review Exercises

1. $\dfrac{7}{8}$
      $-\dfrac{2}{8}$
   _____

2. $\dfrac{3}{4}$
      $+\dfrac{1}{4}$
   _____

3. Convert $\dfrac{14}{10}$ to a mixed numeral.

4. Reduce $\dfrac{16}{20}$ to its lowest terms.

Helpful Hints

Use what you have learned to solve the following problems. Regroup when necessary.

Reduce all answers to their lowest terms. If problems are positioned horizontally, put them in columns before working.

S. $2\dfrac{7}{10}$
      $+5\dfrac{5}{10}$
   _____

S. $7\dfrac{1}{3}$
      $-2\dfrac{2}{3}$
   _____

1. $\dfrac{7}{9}$
      $+\dfrac{3}{9}$
   _____

2. $\dfrac{15}{16}$
      $-\dfrac{3}{16}$
   _____

3. $3\dfrac{3}{5}$
      $+7\dfrac{3}{5}$
   _____

4. 7
      $-2\dfrac{1}{3}$
   _____

5. $6\dfrac{5}{8}$
      $-1\dfrac{1}{8}$
   _____

6. $5-\dfrac{1}{3}=$

7. $3\dfrac{7}{8}$
      $+4\dfrac{5}{8}$
   _____

8. $6\dfrac{1}{3}$
      $-1\dfrac{2}{3}$
   _____

9. $7\dfrac{1}{10}$
      $-3\dfrac{7}{10}$
   _____

10. $\dfrac{3}{5}+\dfrac{4}{5}+\dfrac{3}{5}=$

1	
2	
3	
4	
5	
6	
7	
8	
9	
10	
Score	

Problem Solving

A family has $12\dfrac{1}{3}$ pounds of beef in the freezer. If they used $3\dfrac{2}{3}$ pounds for dinner, how much beef is left?

Fractions

Finding Least Common Denominators

Speed Drills	Review Exercises

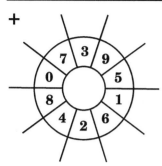

+

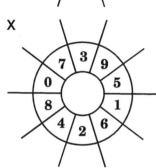

X

1. Find the sum of $\frac{4}{5}$ and $\frac{3}{5}$.

2. Find the difference between $\frac{7}{8}$ and $\frac{3}{8}$.

3.
$$\begin{array}{r} 7 \\ -\ 2\ \frac{3}{4} \\ \hline \end{array}$$

4.
$$\begin{array}{r} 7\ \frac{3}{5} \\ -\ 4 \\ \hline \end{array}$$

Helpful Hints	To add or subtract fractions with unlike denominators you need to find the least common denominator (LCD). The LCD is the smallest number, other than zero, that each denominator will divide into evenly.	**Examples:** The Least Common Denominator of: $\frac{1}{5}$ and $\frac{1}{10}$ is 10 $\frac{3}{8}$ and $\frac{1}{6}$ is 24

Find the least common denominator of each of the following:

S. $\frac{1}{3}$ and $\frac{3}{4}$ S. $\frac{3}{8}$ and $\frac{7}{12}$ 1. $\frac{1}{5}$ and $\frac{4}{15}$ 2. $\frac{5}{6}$ and $\frac{7}{9}$

3. $\frac{9}{14}$ and $\frac{1}{7}$ 4. $\frac{1}{8}, \frac{1}{6}$ and $\frac{1}{4}$ 5. $\frac{5}{9}, \frac{5}{6}$ and $\frac{7}{12}$ 6. $\frac{4}{5}$ and $\frac{1}{4}$

7. $\frac{1}{13}$ and $\frac{7}{39}$ 8. $\frac{5}{12}, \frac{7}{20}$ and $\frac{11}{60}$ 9. $\frac{11}{24}, \frac{3}{16}$ and $\frac{13}{48}$ 10. $\frac{1}{9}, \frac{5}{12}$ and $\frac{5}{6}$

1	
2	
3	
4	
5	
6	
7	
8	
9	
10	
Score	

Problem Solving	A plane traveled 4,500 miles in six hours. What was its average speed per hour?

Fractions   Adding and Subtracting with Unlike Denominators

Speed Drills	Review Exercises

+

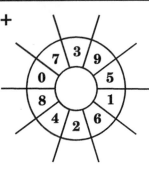

x

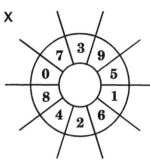

Helpful Hints

To add or subtract fractions with unlike denominators, find the least common denominator. Multiply each fraction by one to make equivalent fractions. Finally, add or subtract.

1. $3\overline{)602}$

2. $\begin{array}{r} 43 \\ \times\ 36 \\ \hline \end{array}$

3. $36 + 7 + 309 =$

4. $600 - 139 =$

Examples:

$$\begin{array}{r} \frac{2}{5} \times \frac{2}{2} = \frac{4}{10} \\ + \frac{1}{2} \times \frac{5}{5} = \frac{5}{10} \\ \hline \frac{9}{10} \end{array}$$

$$\begin{array}{r} \frac{5}{6} \times \frac{2}{2} = \frac{10}{12} \\ + \frac{1}{4} \times \frac{3}{3} = \frac{3}{12} \\ \hline \frac{13}{12} = 1\frac{1}{12} \end{array}$$

S. $\begin{array}{r} \frac{1}{3} \\ + \frac{1}{4} \\ \hline \end{array}$

S. $\begin{array}{r} \frac{4}{5} \\ - \frac{3}{10} \\ \hline \end{array}$

1. $\begin{array}{r} \frac{2}{9} \\ + \frac{1}{3} \\ \hline \end{array}$

2. $\begin{array}{r} \frac{2}{3} \\ - \frac{1}{2} \\ \hline \end{array}$

3. $\begin{array}{r} \frac{5}{6} \\ + \frac{1}{3} \\ \hline \end{array}$

4. $\begin{array}{r} \frac{2}{5} \\ + \frac{1}{3} \\ \hline \end{array}$

5. $\begin{array}{r} \frac{5}{6} \\ - \frac{5}{12} \\ \hline \end{array}$

6. $\begin{array}{r} \frac{1}{2} \\ + \frac{4}{7} \\ \hline \end{array}$

7. $\begin{array}{r} \frac{4}{5} \\ + \frac{7}{10} \\ \hline \end{array}$

8. $\begin{array}{r} \frac{3}{11} \\ + \frac{1}{2} \\ \hline \end{array}$

9. $\begin{array}{r} \frac{4}{7} \\ - \frac{1}{2} \\ \hline \end{array}$

10. $\begin{array}{r} \frac{8}{9} \\ + \frac{1}{4} \\ \hline \end{array}$

1	
2	
3	
4	
5	
6	
7	
8	
9	
10	
Score	

Problem Solving

John bought nine gallons of paint to paint his house. If he used $5\frac{3}{8}$ gallons, how much does he have left?

Fractions

Adding Mixed Numerals with Unlike Denominators

Speed Drills	Review Exercises

Speed Drills

+

x

Helpful Hints

When adding mixed numerals with unlike denominators, first add the fractions. If there is an improper fraction, make it a mixed numeral. Finally, add the sum to the sum of the whole numbers.

*Reduce fractions to lowest terms.

Review Exercises

1. $20 \overline{\smash{)}3762}$

2. $\begin{array}{r} \frac{7}{8} \\ - \frac{1}{8} \\ \hline \end{array}$

3. $\begin{array}{r} \frac{9}{10} \\ + \frac{1}{5} \\ \hline \end{array}$

4. $\begin{array}{r} \frac{3}{4} \\ - \frac{1}{3} \\ \hline \end{array}$

Example:

$$3 \frac{2}{3} \times \frac{2}{2} = \frac{4}{6}$$
$$+ 2 \frac{1}{2} \times \frac{3}{3} = \frac{3}{6}$$
$$\overline{5 \frac{7}{6}} = 1 \frac{1}{6} = 6 \frac{1}{6}$$

S. $\begin{array}{r} 3 \frac{2}{3} \\ + 4 \frac{1}{4} \\ \hline \end{array}$

S. $\begin{array}{r} 4 \frac{1}{2} \\ + 3 \frac{3}{5} \\ \hline \end{array}$

1. $\begin{array}{r} 5 \frac{5}{6} \\ + 2 \frac{1}{3} \\ \hline \end{array}$

2. $\begin{array}{r} 7 \frac{1}{4} \\ + 3 \frac{1}{2} \\ \hline \end{array}$

3. $\begin{array}{r} 5 \frac{7}{8} \\ + 2 \frac{1}{4} \\ \hline \end{array}$

4. $\begin{array}{r} 6 \frac{3}{7} \\ + 2 \frac{1}{14} \\ \hline \end{array}$

5. $\begin{array}{r} 8 \frac{1}{4} \\ + 7 \frac{1}{2} \\ \hline \end{array}$

6. $\begin{array}{r} 7 \frac{3}{10} \\ + 2 \frac{7}{20} \\ \hline \end{array}$

7. $\begin{array}{r} 3 \frac{1}{5} \\ + 2 \frac{1}{10} \\ \hline \end{array}$

8. $\begin{array}{r} 7 \frac{7}{9} \\ + 3 \frac{5}{18} \\ \hline \end{array}$

9. $\begin{array}{r} 6 \frac{1}{3} \\ + 2 \frac{1}{5} \\ \hline \end{array}$

10. $\begin{array}{r} 9 \frac{3}{4} \\ + 2 \frac{1}{6} \\ \hline \end{array}$

1	
2	
3	
4	
5	
6	
7	
8	
9	
10	
Score	

Problem Solving

A factory can produce 352 parts each hour. How many parts can it produce in 12 hours?

Fractions   Subtracting Mixed Numerators with Unlike Denominators

Speed Drills

+

x

Helpful Hints

Review Exercises

1. $\dfrac{3}{7}$

     $+ \dfrac{1}{2}$

2. $5\dfrac{7}{8}$

    $+ 4\dfrac{1}{8}$

3. 3

    $+ 1\dfrac{2}{5}$

4. $4\dfrac{1}{3}$

    $- 2\dfrac{2}{3}$

To subtract mixed numerators with unlike denominators, first subtract the fractions. If the fractions cannot be subtracted, take one from the whole number, increase the fraction, then subtract.

Examples:

$$\overset{5}{\cancel{6}}\dfrac{1}{6} = \dfrac{2}{12} + \dfrac{12}{12} = \dfrac{14}{12}$$
$$- 3\dfrac{1}{4} = \dfrac{3}{12}$$
$$2\dfrac{11}{12}$$

$$7\dfrac{1}{2} \times \dfrac{3}{3} = \dfrac{3}{6}$$
$$- 2\dfrac{1}{3} \times \dfrac{2}{2} = \dfrac{2}{6}$$
$$5 \qquad \dfrac{1}{6} = 5\dfrac{1}{6}$$

S. $4\dfrac{1}{4}$ S. $5\dfrac{1}{3}$ 1. $3\dfrac{7}{8}$ 2. $9\dfrac{5}{6}$

    $- 1\dfrac{1}{5}$ $- 2\dfrac{1}{2}$ $- 1\dfrac{1}{4}$ $- 2\dfrac{1}{3}$

3. $5\dfrac{1}{4}$ 4. $7\dfrac{1}{8}$ 5. $2\dfrac{1}{7}$ 6. $9\dfrac{1}{4}$

    $- 2\dfrac{2}{3}$ $- 2\dfrac{1}{2}$ $- 1\dfrac{3}{14}$ $- 3\dfrac{7}{16}$

7. $6\dfrac{2}{3}$ 8. $6\dfrac{1}{2}$ 9. $7\dfrac{1}{4}$ 10. $7\dfrac{1}{8}$

    $- 3\dfrac{4}{9}$ $- 2\dfrac{2}{3}$ $- 2\dfrac{3}{5}$ $- 4\dfrac{3}{4}$

1	
2	
3	
4	
5	
6	
7	
8	
9	
10	
Score	

Problem Solving

There are 312 students enrolled in a school. If they have been placed into thirteen equal-sized classes, how many students are in each class?

Speed Drills	Review Exercises

+

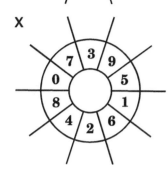

X

1. Reduce $\dfrac{24}{30}$ to its lowest terms

2. Convert $\dfrac{29}{4}$ to a mixed numeral

3. Find the least common denominator for the following fractions:
$\dfrac{1}{3}$, $\dfrac{5}{6}$, and $\dfrac{3}{4}$

4.
$$\begin{array}{r} \frac{1}{3} \\ \frac{1}{5} \\ +\ \frac{1}{2} \\ \hline \end{array}$$

Helpful Hints	Use what you have learned to solve the following problems. *Be sure all fractions are reduced to lowest terms.

S. $3\frac{1}{7}$ $-1\frac{5}{7}$

S. $6\frac{1}{2}$ $+3\frac{3}{4}$

1. $\frac{8}{9}$ $-\frac{1}{2}$

2. $\frac{8}{9}$ $-\frac{1}{6}$

3. 7 $-2\frac{3}{5}$

4. $5\frac{1}{8}$ $+3\frac{1}{2}$

5. $7\frac{7}{8}$ $+3\frac{3}{8}$

6. $7\frac{1}{2}$ $-2\frac{3}{4}$

7. $3\frac{4}{5}$ $+4\frac{2}{3}$

8. $6\frac{1}{2}$ -3

9. $\frac{7}{16}$ $+\frac{1}{4}$

10. $4\frac{5}{6}$ $+3\frac{3}{4}$

1	
2	
3	
4	
5	
6	
7	
8	
9	
10	
Score	

Problem Solving	Susan earned $3\frac{3}{4}$ dollars on Monday and $7\frac{1}{2}$ dollars on Tuesday. How much more did she earn on Tuesday than on Monday?

Speed Drills

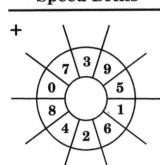

+

x

Review Exercises

1. $6\overline{)726}$

2.
$$725$$
$$\times\ 36$$

3. $73 + 13 + 76 + 59 =$

4. $8,033 - 1,765 =$

Helpful Hints	When multiplying common fractions, first multiply the numerators. Next, multiply the denominators. If the answer is an improper fraction, change it to a mixed numeral.	**Examples:** $\frac{3}{4} \times \frac{2}{7} = \frac{6}{28} = \frac{3}{14}$ $\frac{3}{2} \times \frac{7}{8} = \frac{21}{16} = 1\frac{5}{16}$	*Be sure to reduce fractions to lowest terms.

S. $\frac{3}{4} \times \frac{5}{7} = \frac{15}{28}$ S. $\frac{4}{5} \times \frac{3}{5} =$ 1. $\frac{2}{9} \times \frac{1}{7} =$ 2. $\frac{2}{5} \times \frac{1}{2} =$

3. $\frac{7}{2} \times \frac{3}{5} =$ 4. $\frac{7}{9} \times \frac{2}{3} =$ 5. $\frac{8}{9} \times \frac{3}{4} =$ 6. $\frac{4}{3} \times \frac{4}{5} =$

7. $\frac{3}{2} \times \frac{4}{5} =$ 8. $\frac{2}{7} \times \frac{3}{5} =$ 9. $\frac{1}{2} \times \frac{4}{5} =$ 10. $\frac{5}{2} \times \frac{3}{7} =$

1	
2	
3	
4	
5	
6	
7	
8	
9	
10	
Score	

Problem Solving

A family roasted $2\frac{1}{4}$ pounds of beef for dinner and ate $1\frac{3}{5}$ pounds. How much beef was left?

Fractions · Multiplying Common Fractions/Eliminating Common Factors

Speed Drills	Review Exercises

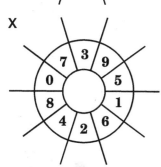

+

X

1. $\dfrac{2}{3} \times \dfrac{6}{7} =$

2. $\dfrac{7}{3} \times \dfrac{4}{5} =$

3. $\dfrac{7}{8}$
    $-\dfrac{1}{8}$
    ___

4. $\dfrac{2}{3}$
    $+\dfrac{2}{3}$
    ___

Helpful Hints	If the numerator of one fraction and the denominator of another have a common factor, they can be divided out before you multiply the fractions.	**Examples:** 4 is a common factor $\dfrac{3}{1\cancel{4}} \times \dfrac{\cancel{8}^2}{11} = \dfrac{6}{11}$	2 is a common factor $\dfrac{7}{4\cancel{8}} \times \dfrac{\cancel{6}^3}{5} = \dfrac{21}{20} = 1\dfrac{1}{20}$

S. $\dfrac{3}{5} \times \dfrac{5}{7} =$ S. $\dfrac{9}{10} \times \dfrac{5}{3} =$ 1. $\dfrac{2}{5} \times \dfrac{15}{16} =$ 2. $\dfrac{8}{15} \times \dfrac{3}{16} =$

3. $\dfrac{5}{6} \times \dfrac{7}{15} =$ 4. $\dfrac{7}{3} \times \dfrac{10}{7} =$ 5. $\dfrac{5}{8} \times \dfrac{12}{25} =$ 6. $\dfrac{8}{9} \times \dfrac{3}{4} =$

7. $\dfrac{3}{4} \times \dfrac{8}{15} =$ 8. $\dfrac{3}{4} \times \dfrac{3}{5} =$ 9. $\dfrac{4}{3} \times \dfrac{6}{7} =$ 10. $\dfrac{5}{6} \times \dfrac{4}{7} =$

1	
2	
3	
4	
5	
6	
7	
8	
9	
10	
Score	

Problem Solving	There are 15 rows of seats in a theater. If each row has 26 seats, how many seats are there in the theater?

| **Speed Drills** | **Review Exercises** |

+

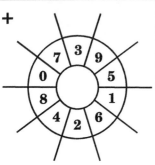

1. $\dfrac{3}{5} \times \dfrac{15}{21} =$ 2. $\dfrac{8}{9} \times \dfrac{7}{12} =$

X

3. $\dfrac{2}{3}$ 4. $\dfrac{3}{4}$

     $+ \dfrac{1}{5}$ $- \dfrac{2}{3}$
     _____   _____

| **Helpful Hints** | When multiplying whole numbers and fractions, write the whole number as a fraction and then multiply. | **Examples:** $\dfrac{2}{3} \times 15 =$ $\dfrac{2}{1\cancel{3}} \times \dfrac{\cancel{15}^{5}}{1} = \dfrac{10}{1} = 10$ | $\dfrac{3}{4} \times 9 =$ $\dfrac{3}{4} \times \dfrac{9}{1} = \dfrac{27}{4}$ | $\begin{array}{r} 6\frac{3}{4} \\ 4\overline{)27} \\ 24 \\ \hline 3 \end{array}$ |

S. $\dfrac{3}{4} \times 12 =$ S. $\dfrac{2}{3} \times 5 =$ 1. $\dfrac{3}{4} \times 8 =$ 2. $10 \times \dfrac{2}{5} =$

1	
2	
3	

3. $\dfrac{4}{5} \times 25 =$ 4. $\dfrac{2}{7} \times 4 =$ 5. $\dfrac{1}{2} \times 27 =$ 6. $\dfrac{1}{10} \times 25 =$

4	
5	
6	
7	

7. $6 \times \dfrac{7}{12} =$ 8. $\dfrac{5}{6} \times 9 =$ 9. $\dfrac{3}{8} \times 40 =$ 10. $\dfrac{4}{5} \times 7 =$

8	
9	
10	
Score	

| **Problem Solving** | A class has 36 students. If $\frac{2}{3}$ of them are girls, how many girls are there in the class? |

Fractions

Speed Drills	**Review Exercises**

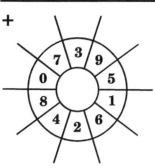

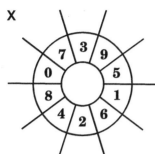

1. $63\overline{)796}$

2. $\dfrac{3}{4} \times 16 =$

3. $\dfrac{2}{3} \times 10 =$

4. Change $3\dfrac{1}{2}$ to an improper fraction.

Helpful Hints	To multiply mixed numerals, first change them to improper fractions, then multiply. Express answers in lowest terms.	**Example:** $1\dfrac{1}{2} \times 1\dfrac{5}{6} =$ $\dfrac{1\cancel{3}}{2} \times \dfrac{11}{\cancel{6}_2} = \dfrac{11}{4} = 2\dfrac{3}{4}$

S. $\dfrac{2}{3} \times 1\dfrac{1}{8} =$ S. $1\dfrac{1}{4} \times 2\dfrac{2}{5} =$ 1. $\dfrac{3}{4} \times 2\dfrac{1}{2} =$ 2. $2\dfrac{1}{3} \times 1\dfrac{1}{3} =$

3. $5 \times 3\dfrac{1}{5} =$ 4. $2\dfrac{1}{7} \times 1\dfrac{2}{5} =$ 5. $2\dfrac{2}{3} \times 2\dfrac{1}{4} =$ 6. $2\dfrac{1}{4} \times 1\dfrac{1}{2} =$

7. $2\dfrac{1}{2} \times 3\dfrac{1}{4} =$ 8. $6 \times 2\dfrac{1}{2} =$ 9. $2\dfrac{1}{2} \times 4\dfrac{2}{3} =$ 10. $2\dfrac{1}{6} \times \dfrac{3}{5} =$

1	
2	
3	
4	
5	
6	
7	
8	
9	
10	
Score	

Problem Solving	If a man can run 4 miles in an hour, at this rate, how far can he run in $3\dfrac{1}{2}$ hours?

Fractions

Speed Drills	Review Exercises

+

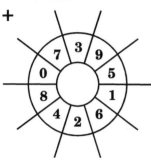

x

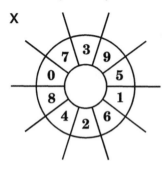

1. $\dfrac{3}{5}$ x $\dfrac{4}{7}$ =

2. $\dfrac{3}{4}$ x $\dfrac{9}{25}$ =

3. $\dfrac{3}{4}$ x 24 =

4. 13 x $\dfrac{2}{3}$ =

Helpful Hints	Use what you have learned to solve the following problems.	*Be sure to express answers in lowest terms. *Sometimes common factors may be divided out before you multiply.

S. $\dfrac{3}{5}$ x $\dfrac{1}{2}$ =

S. $3\dfrac{1}{2}$ x $2\dfrac{1}{7}$ =

1. $\dfrac{4}{5}$ x $\dfrac{7}{8}$ =

2. $\dfrac{20}{21}$ x $\dfrac{7}{40}$ =

3. $\dfrac{3}{5}$ x 35 =

4. $\dfrac{4}{7}$ x 9 =

5. $\dfrac{3}{4}$ x $2\dfrac{1}{2}$ =

6. $3\dfrac{2}{3}$ x $\dfrac{1}{2}$ =

7. 5 x $3\dfrac{2}{5}$ =

8. $1\dfrac{2}{3}$ x $1\dfrac{2}{5}$ =

9. $3\dfrac{1}{6}$ x $4\dfrac{4}{5}$ =

10. $1\dfrac{7}{8}$ x $4\dfrac{1}{3}$ =

1	
2	
3	
4	
5	
6	
7	
8	
9	
10	
Score	

Problem Solving	If a factory can produce $4\dfrac{1}{2}$ tons of parts in a day, how many tons can it produce in 5 days?

Fractions Reciprocals

Speed Drills	Review Exercises

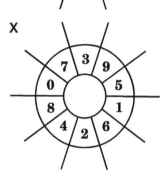

+

x

1. $\dfrac{3}{5}$
    $+\dfrac{1}{3}$
    _____

2. $\dfrac{3}{4}$
    $-\dfrac{1}{2}$
    _____

3. $2 \times 3\dfrac{1}{2} =$

4. $1\dfrac{1}{3} \times 1\dfrac{1}{3} =$

Helpful Hints

To find the reciprocal of a common fraction, invert the fraction. To find the reciprocal of a mixed numeral, change the mixed numeral to an improper fraction, then invert it. To find the reciprocal of a whole number, first make it a fraction then invert it.

Examples: The reciprocal of:

$\dfrac{3}{5}$ is $\dfrac{5}{3}$ or $1\dfrac{2}{3}$ $2\dfrac{1}{2}$ is $\dfrac{2}{5}$ 7 is $\dfrac{1}{7}$
$\left(\dfrac{5}{2}\right)$ $\left(\dfrac{7}{1}\right)$

Find the reciprocal of each number:

S. $\dfrac{3}{4}$ S. $2\dfrac{1}{3}$ 1. 6   2. $\dfrac{7}{8}$

3. $3\dfrac{1}{4}$ 4. 13   5. $\dfrac{2}{5}$ 6. $\dfrac{1}{7}$

7. 9   8. $\dfrac{2}{9}$ 9. $4\dfrac{1}{2}$ 10. 15

1	
2	
3	
4	
5	
6	
7	
8	
9	
10	
Score	

Problem Solving

Five students earned 225 dollars. If they divided the money equally among themselves, how much did each student receive?

Fractions   Dividing Fractions and Mixed Numbers

Speed Drills	Review Exercises

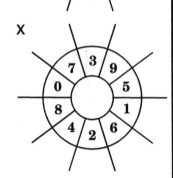

1. Find the reciprocal of 9

2. Find the reciprocal of

$$\frac{2}{7}$$

3. Find the reciprocal of

$$3 \frac{2}{3}$$

4. $2\frac{2}{3}$ x $1\frac{1}{5}$ =

Helpful Hints

To divide fractions, find the reciprocal of the second number, then multiply the fractions.

Examples:

$$\frac{2}{3} \div \frac{1}{2} =$$

$$\frac{2}{3} \text{ x } \frac{2}{1} = \frac{4}{3} = \boxed{1\frac{1}{3}}$$

$$2\frac{1}{2} \div 1\frac{1}{2} = \frac{5}{2} \div \frac{3}{2} =$$

$$\frac{5}{2} \text{ x } \frac{2}{3} = \frac{5}{3} = \boxed{1\frac{2}{3}}$$

S. $\frac{3}{7} \div \frac{3}{8}$ =

S. $3\frac{1}{2} \div 2$ =

1. $\frac{3}{8} \div \frac{1}{6}$ =

2. $\frac{1}{2} \div \frac{1}{3}$ =

3. $5 \div \frac{2}{3}$ =

4. $4\frac{1}{2} \div \frac{1}{2}$ =

5. $1\frac{3}{4} \div \frac{3}{8}$ =

6. $5\frac{1}{4} \div \frac{7}{12}$ =

7. $1\frac{1}{2} \div 3$ =

8. $5\frac{1}{2} \div 2$ =

9. $7\frac{1}{2} \div 2\frac{1}{2}$ =

10. $3\frac{2}{3} \div 2\frac{1}{2}$ =

1	
2	
3	
4	
5	
6	
7	
8	
9	
10	
Score	

Problem Solving

$3\frac{1}{2}$ yards are to be divided into pieces that are $\frac{1}{2}$ yards long. How many pieces will there be?

Reviewing All Fraction Operations

1.
$$\frac{3}{5}$$
$$+\frac{1}{5}$$

2.
$$\frac{5}{6}$$
$$+\frac{3}{6}$$

3.
$$\frac{2}{3}$$
$$+\frac{1}{5}$$

4.
$$3\frac{2}{3}$$
$$+4\frac{5}{9}$$

5.
$$7\frac{3}{4}$$
$$+2\frac{3}{8}$$

6.
$$\frac{5}{8}$$
$$-\frac{1}{8}$$

7.
$$6\,\backslash 7\frac{2}{5}\!+\!\frac{5}{5}\!=\!\frac{7}{5}$$
$$-2\frac{3}{5}$$

8.
$$7$$
$$-2\frac{3}{5}$$

9.
$$6\frac{3}{4}$$
$$-\frac{1}{2}$$

10.
$$9\,^{+}\frac{1}{3}$$
$$-3\,^{+}\frac{2}{5}$$

11. $\frac{2}{3}$ x $\frac{4}{7}$ =

12. $\frac{12}{13}$ x $\frac{3}{24}$ =

13. $\left(\frac{3}{4}\text{ x }36 = \right)$

14. $\frac{7}{8}$ x $2\frac{1}{7}$ =

15. $2\frac{1}{3}$ x $3\frac{1}{2}$ =

16. $\frac{3}{4}$ ÷ $\frac{1}{2}$ =

17. $3\frac{1}{2}$ ÷ $\frac{1}{2}$ =

18. $3\frac{2}{3}$ ÷ $1\frac{1}{2}$ =

19. $3\frac{3}{4}$ ÷ $1\frac{1}{8}$ =

20. $6 \div 2\frac{1}{3}$ =

1	
2	
3	
4	
5	
6	
7	
8	
9	
10	
11	
12	
13	
14	
15	
16	
17	
18	
19	
20	

Speed Drills	Review Exercises

+

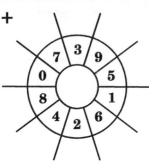

x

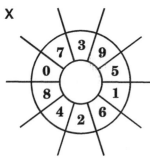

1.　136 + 927 + 813

2.　$\dfrac{3}{5}$ $+ \dfrac{2}{3}$

2.　$\begin{array}{r} 1{,}394 \\ -\ \ 966 \end{array}$

4.　$\dfrac{3}{4}$ $-\dfrac{1}{6}$

Helpful Hints

ones　tenths　hundredths　thousandths　ten-thousandths　hundred-thousandths　millionths

1 . 2 3 4 5 6 7

To read decimals first read the whole number. Next, read the decimal point as "and." Next, read the number after the decimal point and its place value.

Examples:
3.16 = three and sixteen hundredths
14.011 = fourteen and eleven thousandths
0.69 = sixty-nine hundredths

Write the following in words:

S.　2.6　　　S.　13.016　　　1.　0.73　　　2.　4.002

3.　132.6　　　4.　132.06　　　5.　72.6395　　　6.　0.077

7.　9.89　　　8.　6.003　　　9.　0.72　　　10.　1.666

1	
2	
3	
4	
5	
6	
7	
8	
9	
10	
Score	

Problem Solving　　In a class of 35 students, $\frac{3}{5}$ of them are boys. How many boys are there in the class?

Speed Drills	**Review Exercises**

+

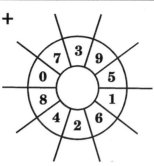

x

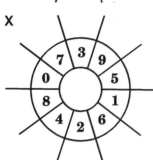

1. 234
    x 36

2. $\dfrac{3}{4}$ x $\dfrac{8}{11}$ =

3. $2\dfrac{1}{2} \div \dfrac{1}{2}$ =

4. $3\dfrac{1}{3} \div 2$ =

Helpful Hints

When reading remember "and" means decimal point. The fraction part of a decimal ends in "th" or "ths." Be careful about placeholders.

Examples:
Four and eight tenths = 4.8
Two hundred one and six hundredths = 201.06
One hundred four ten-thousandths = .0104

Write each of the following as a decimal. Use the chart at the bottom to help.

S. Six and four hundredths

S. Three hundred six and fifteen hundredths

1. Nine and eight tenths

2. Forty-six and thirteen thousandths

3. Three hundred twenty-six ten-thousandths

4. Fifty and thirty-nine thousandths

5. Eight hundred-thousandths

6. Four millionths

7. Twelve and thirty-six ten thousandths

8. Sixteen and twenty-four thousandths

9. Twenty-three and five tenths

10. Two and seventeen thousandths

1	
2	
3	
4	
5	
6	
7	
8	
9	
10	
Score	

Problem Solving

If a normal temperature is $98\dfrac{3}{5}$ degrees, and a man has a temperature of 102 degrees, how much above normal is his temperature?

Speed Drills	**Review Exercises**

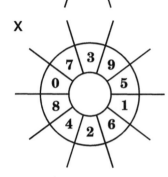

+

X

1. $\dfrac{3}{4} \div \dfrac{1}{2} =$ 2. $1\dfrac{1}{2} \div 2 =$

3. $3 \div 1\dfrac{1}{2} =$ 4. $3\dfrac{1}{3} \div \dfrac{2}{5} =$

Helpful Hints

When changing mixed numerals to decimals, remember to put a decimal point after the whole number.

Examples: $3\dfrac{3}{10} = 3.3$ $\dfrac{16}{10,000} = .0016$

$42\dfrac{9}{10,000} = 42.0009$ $65\dfrac{12}{100,000} = 65.00012$

Write each of the following as a decimal. Use the chart at the bottom to help.

S. $7\dfrac{7}{10}$ S. $9\dfrac{7}{1,000}$ 1. $16\dfrac{32}{100}$ 2. $\dfrac{97}{10,000}$

3. $72\dfrac{9}{100}$ 4. $134\dfrac{92}{10,000}$ 5. $\dfrac{16}{1,000}$ 6. $44\dfrac{432}{100,000}$

7. $3\dfrac{96}{1,000}$ 8. $4\dfrac{901}{1,000}$ 9. $3\dfrac{901}{1,000,000}$ 10. $\dfrac{1,763}{100,000}$

1	
2	
3	
4	
5	
6	
7	
8	
9	
10	
Score	

ones   tenths   hundredths   thousandths   ten-thousandths   hundred-thousandths   millionths

9 . 8 7 6 5 4 3

Problem Solving

A floor tile is $\dfrac{3}{4}$ inches thick. How many inches thick would a stack of forty-eight tiles be?

Decimals | Changing Decimals to Fractions and Mixed Numerals

Speed Drills	**Review Exercises**

+

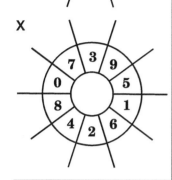

x

1.
$$\frac{2}{3}$$
$$+ \frac{1}{5}$$
_____

2.
$$7$$
$$- 1\frac{1}{3}$$
_____

3. $\frac{2}{3} \times 1\frac{1}{2} =$

4. $\frac{2}{3} \div 5\frac{1}{2} =$

Helpful Hints	Decimals can be easily changed to mixed numerals and fractions. Remember that the whole number is to the left of the decimal.	**Examples:** $2.6 = 2\frac{6}{10}$ $\quad .210 = \frac{210}{1,000}$ $3.007 = 3\frac{7}{1,000}$ $\quad 1.0019 = 1\frac{19}{10,000}$

Change each of the following to a mixed numeral or fraction. Use the chart at bottom for help.

S. 1.43	S. 7.006	1. 173.016	2. .00016
3. 7.000014	4. 19.936	5. .09163	6. 77.8
7. 13.019	8. 72.0009	9. .00099	10. 63.000143

ones . tenths hundredths thousandths ten-thousandths hundred-thousandths millionths

9 . 8 7 6 5 4 3

1	
2	
3	
4	
5	
6	
7	
8	
9	
10	
Score	

Problem Solving	A theater has 9 rows of fourteen seats. Six of the seats are empty. How many of the seats are taken?

Speed Drills	**Review Exercises**

+

1. Reduce $\frac{25}{30}$ to its simplest form.

2. Change $\frac{35}{8}$ to a mixed numeral.

X

3. Write 1.019 in words.

4. Change 72.008 to a mixed numeral.

Helpful Hints	Zeroes can be put to the right of a decimal without changing the value. This helps when comparing the value of decimals.	**< means less than** **> means greater than**	**Example:** Compare 4.3 and 4.28 4.3 =4.30 so 4.3 > 4.28

Place > or < to compare each pair of decimals.

S. 7.32 ☐ 7.6   S. .99 ☐ .987   1. 6.096 ☐ 6.1

2. 3.41 ☐ 3.336   3. 7.11 ☐ 7.09   4. 1.5 ☐ 1.42

5. 3.62 ☐ .099   6. .6 ☐ .79   7. 2.31 ☐ 2.4

8. 1.64 ☐ 1.596   9. 3.09 ☐ 3.4   10. 6.199 ☐ 6.2

1	
2	
3	
4	
5	
6	
7	
8	
9	
10	
Score	

Problem Solving	A group of hikers needed to travel 43 miles. They traveled 7 miles per day. After 5 days, how much farther did they still have to hike?

Decimals

Speed Drills	Review Exercises

+

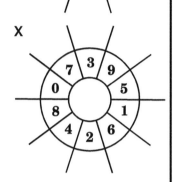

X

1. $3\frac{1}{2} \div 2 =$ 2. $2\frac{1}{2} \div 2 =$

3. $\begin{array}{r} \frac{3}{8} \\ + \frac{5}{8} \\ \hline \end{array}$ 4. $\begin{array}{r} 7\frac{3}{5} \\ + 6\frac{4}{5} \\ \hline \end{array}$

Helpful Hints	To add decimals, line up the decimal points and add as you would whole numbers. Write the decimal points in the answer. Zeroes may be placed to the right of the decimal.	**Example:** Add 3.16 + 2.4 + 12 $\begin{array}{r} 3.16 \\ 2.40 \\ + 12.00 \\ \hline 17.56 \end{array}$

S. $\begin{array}{r} 3.16 \\ 12.4 \\ + 3.26 \\ \hline \end{array}$

S. 3.92 + 4.6 + .32 =

1. 32.16 + 1.7 + 7.493 =

2. 7.341 + 6.49 + .6 =

3. $\begin{array}{r} 7.64 \\ 19.633 \\ + 2.4 \\ \hline \end{array}$

4. .37 + .6 + .73 =

5. 9.64 + 7 + 1.92 + .7 =

6. 72.163 + 11.4 + 63.42 =

7. .7 + .6 + .4 =

8. 17.33 + 6.994 + .72 =

9. $\begin{array}{r} 7.642 \\ 17.63 \\ 2.143 \\ + 14.64 \\ \hline \end{array}$

10. 19.2 + 7.63 + 4.26 =

1	
2	
3	
4	
5	
6	
7	
8	
9	
10	
Score	

Problem Solving	In January it rained 3.6 inches, in February, 4.3 inches, and in March, 7.9 inches. What was the total amount of rainfall for the three months?

Speed Drills	Review Exercises

+

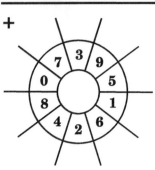

x

1.   3.16
                    3.4
             + 7.166

2. $3.6 + 4.16 + 8 =$

3. Find $\frac{1}{2}$ of $3\frac{1}{2}$

4. Write $72\frac{9}{1,000}$ as a decimal.

Helpful Hints	To subtract decimals, line up the decimal points and subtract as you would whole numbers. Write the decimal points in the answer. Zeroes may be placed to the right of the decimal.	**Examples:**

$$3.2 - 1.66 = \quad\quad 7 - 1.63 =$$

$$\overset{2\ \ 11\ 1}{\cancel{3}.\cancel{2}0} \quad\quad\quad \overset{6\ \ 9\ 1}{\cancel{7}.\cancel{0}0}$$
$$\underline{-\ 1.66} \quad\quad\quad \underline{-\ 1.63}$$
$$1.54 \quad\quad\quad\quad 5.37$$

S.   17.2
     − 3.36
	S. $15.1 - 7.62 =$	1. 7.32
		     − 1.426
			1	

			2	
2. 3.962 − 1.669	3. $2.72 - 1.56 =$	4. $27.93 - 16.8 =$	**3**	
			4	
			5	
5. $.72 - .667 =$	6. 6.137 − 2.1793	7. $3 - .627 =$	**6**	
			7	
8. $7.14 - 3.456 =$	9. $75.6 - 66.972 =$	10. $43.21 - 16.445 =$	**8**	
			9	
			10	
			Score	

Problem Solving	Bill ran a race in 17.6 seconds and Jane ran it in 16.3 seconds. How much faster did Jane run the race than Bill?

Decimals

Speed Drills	Review Exercises

+

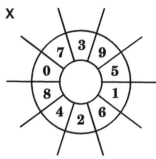

x

1. $\frac{1}{3}$
$+ \frac{1}{3}$
_____

2. $\frac{3}{8}$
$+ \frac{3}{8}$
_____

3. $\frac{7}{16}$
$+ \frac{13}{16}$
_____

4. $\frac{4}{5}$
$\frac{3}{5}$
$+ \frac{4}{5}$
_____

Helpful Hints	Use what you have learned to solve the following problems.	* Line up the decimal points. * Put decimal point in the answer. * Zeroes may be added to the right of the decimal point.

S. 3.61
   14.4
 + .37
 _____

S. 7.16
 — 3.473
 _____

1. 7.16
   8.92
 + 7.634
 _____

2. 7.6
 — 1.43
 _____

3. 4.63 + 5.7 + 6.24 =

4. 17.2 — 8.96 =

5. 15 — 12.92 =

6. 6.93 + 5 + 7.63 =

7. .9 + .7 + .6 =

8. 7.16 — 2.673 =

9. 27.16 — 16.764 =

10. 7.73 + 2.6 + .37 + 15 =

1	
2	
3	
4	
5	
6	
7	
8	
9	
10	
Score	

Problem Solving	A worker earned $125.65. If he spent $76.93, how much of his earnings was left?

Speed Drills	Review Exercises

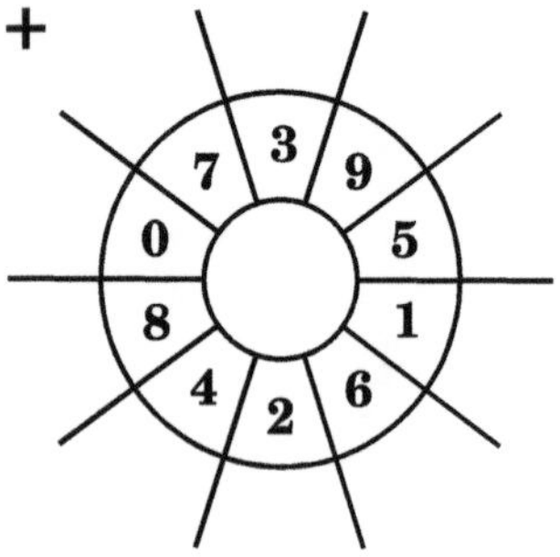

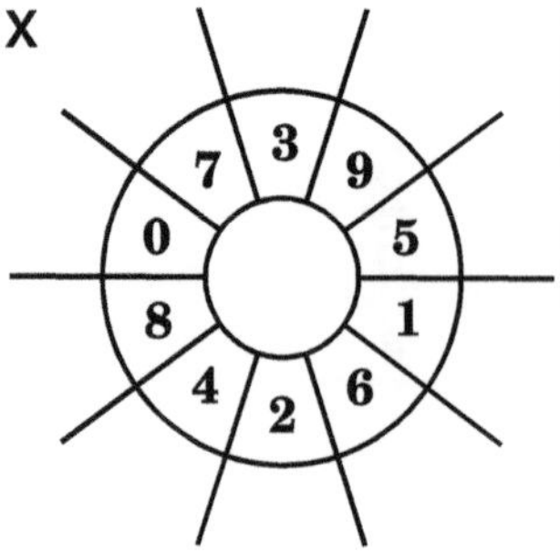

+

X

Helpful Hints

1. 36
    x 6

2. 46
    x 32

3. 209
    x 23

4. 434
    x 612

Multiply as you would with whole numbers. Find the number of decimal places and place the decimal point properly in the product.

Examples:

2.32 ← 2 places
x 6
13.92 ← 2 places

7.6 ← 1 place
x 23
228
1520
174.8 ← 1 place

S.	2.46 x 3	S.	2.3 x 16	1.	.643 x 3	2.	3.66 x 4		

1	
2	
3	
4	
5	
6	
7	
8	
9	
10	
Score	

3. .16
    x 43

4. .236
    x 24

5. 1.4
    x 16

6. 3.45
    x 16

7. 7.63
    x 43

8. 1.432
    x 7

9. .41
    x 73

10. .046
    x 27

Problem Solving

A hiker can travel 2.7 miles in an hour. At this pace, how far can the hiker travel in seven hours?

Speed Drills	Review Exercises

+

X

1.   72.4
     x   3

2.   .27
        x   16

3. $\dfrac{3}{4}$ x $1\dfrac{1}{2}$ =

4. $2\dfrac{1}{8} \div 2$ =

Helpful Hints

Multiply as you would with whole numbers. Find the number of decimal places and place the decimal point properly in the product.

Examples:

$2.63 \leftarrow 2$ places
$\underline{x\ \ .3} \leftarrow 1$ place
$.789 \leftarrow 3$ places

$.724 \leftarrow 3$ places
$\underline{x\ .23} \leftarrow 2$ places
2172
$\underline{14480}$
$.16652 \leftarrow 5$ places

S. 3.6 x .7	S. 3.24 x 2.4	1. 3.6 x 3.2	2. 2.09 x .22
3. .642 x .33	4. .23 x 3.8	5. 2.03 x .07	6. .422 x 23.2
7. .003 x 0.8	8. 5.6 x .34	9. 63.5 x 2.35	10. 12.3 x .006

1	
2	
3	
4	
5	
6	
7	
8	
9	
10	
Score	

Problem Solving

A farmer can harvest 2.5 tons of potatoes in a day. How many tons can be harvested in 4.5 days?

Speed Drills	**Review Exercises**

+

1. 6

   $- 2\dfrac{1}{3}$

2. $3\dfrac{1}{4}$

   $- 1\dfrac{3}{4}$

x

3. $\dfrac{2}{3}$

   $+ \dfrac{2}{3}$

4. $3\dfrac{1}{2}$

   $+ 4\dfrac{1}{2}$

Helpful Hints	To multiply by 10, move the decimal point one place to the right; by 100, two places to the right; by 1,000, three places to the right.	**Examples:** 10 x 3.36 = 33.6 100 x 3.36 = 336 1,000 x 3.36 = 3360* *Sometimes placeholders are necessary

S.   10 x 3.2 =

S.   1,000 x 7.39 =

1.   100 x .936 =

2.   1,000 x 72.6 =

3.   100 x 1.6 =

4.   7.362 x 100 =

5.   7.28 x 1,000 =

6.   100 x .7 =

7.   100 x .376 =

8.   1,000
     x .39

9.   100 x .733 =

10.   10 x 7.63 =

1	
2	
3	
4	
5	
6	
7	
8	
9	
10	
Score	

Problem Solving	If tickets to a concert cost $9.50, how much would 1,000 tickets cost?

Speed Drills	Review Exercises

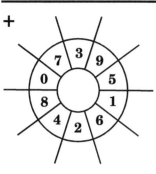

+

x

1. $2\dfrac{1}{2} \div \dfrac{1}{3} =$ 2. $3 \times 2\dfrac{1}{3} =$

3. $\dfrac{16}{17} \times \dfrac{7}{8} =$ 4. $5 \div \dfrac{1}{2} =$

Helpful Hints	Use what you have learned to solve the following problems. *Be careful when placing the decimal point in the product.

S.   .342
     x 7

S.   42.3
     x .36

1.   .23
     x 14

2.   .29
    x 1.6

3.   1.34
    x .362

4. 10 x 2.6 =

5.   2.63
    x 1.2

6. 100 x 26.3 =

7.   .003
    x 3.6

8.   .65
    x 5.5

9.   1.67
    x 33

10.   0.67
    x .063

1	
2	
3	
4	
5	
6	
7	
8	
9	
10	
Score	

Problem Solving	Sweaters are on sale for $13.79. How much would three of the sweaters cost on sale?

Decimals

Speed Drills

Review Exercises

+

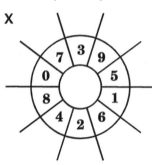

X

Helpful Hints

1. $7 \overline{)133}$

2. $7 \overline{)1414}$

3. $6 \overline{)6006}$

4. $22 \overline{)2442}$

Divide as you would with whole numbers. Place the decimal point directly up.

Examples:

$$3 \overline{)\begin{array}{l} 2.8 \\ 8.4 \\ \underline{-6\downarrow} \\ 24 \\ \underline{-24} \\ 0 \end{array}}$$

$$3 \overline{)\begin{array}{l} .084 \\ .252 \\ \underline{-24\downarrow} \\ 12 \\ \underline{-12} \\ 0 \end{array}}$$

S. $3 \overline{)1.32}$ S. $8 \overline{)14.4}$ 1. $3 \overline{)59.1}$ 2. $7 \overline{)22.47}$

3. $34 \overline{)19.38}$ 4. $70.3 \div 19 =$ 5. $4 \overline{)24.32}$ 6. $6 \overline{)245.4}$

7. $26 \overline{)8.424}$ 8. $16 \overline{)2.56}$ 9. $12.72 \div 6 =$ 10. $21 \overline{)42.84}$

1	
2	
3	
4	
5	
6	
7	
8	
9	
10	
Score	

Problem Solving

A man has a board $10\frac{1}{2}$ feet long. If he decides to cut it into pieces $\frac{1}{2}$ foot long, how many pieces will there be?

Decimals   Dividing Decimals by Whole Numbers with Placeholders

Speed Drills	Review Exercises

+

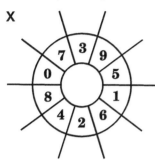

X

1. $3\overline{)6.54}$

2. $15\overline{)4.5}$

3. $3.63 + 12 + 3.2 =$

4. $\begin{array}{r} 7.2 \\ -\ 2.367 \\ \hline \end{array}$

Helpful Hints	Sometimes placeholders are necessary when dividing decimals.	**Examples:** $\begin{array}{r} .05 \\ 3\overline{)\ .15} \\ -15 \\ \hline 0 \end{array}$ $\begin{array}{r} .003 \\ 15\overline{)\ .045} \\ -\ 45 \\ \hline 0 \end{array}$

S. $5\overline{)\ .0135}$ S. $13\overline{)\ .247}$ 1. $7\overline{)\ .0049}$ 2. $3\overline{)\ .036}$

3. $4\overline{)\ .224}$ 4. $13\overline{)\ .468}$ 5. $22\overline{)\ .946}$ 6. $9\overline{)\ .567}$

7. $52\overline{)\ 1.196}$ 8. $18\overline{)\ .396}$ 9. $9\overline{)\ .027}$ 10. $12\overline{)\ .816}$

1	
2	
3	
4	
5	
6	
7	
8	
9	
10	
Score	

Problem Solving	Five boys earned $27.55 doing yard work. If they decided to divide the money equally among themselves, how much would each receive?

Decimals

Speed Drills	Review Exercises

+

X

Helpful Hints

1. $\begin{array}{r} 11.4 \\ \times\ \ 3 \\ \hline \end{array}$

2. $\begin{array}{r} .233 \\ \times\ \ .4 \\ \hline \end{array}$

3. $4\dfrac{1}{2} \div 1\dfrac{1}{2} =$

4. $\dfrac{2}{3} \div \dfrac{2}{9} =$

Sometimes zeroes need to be added to the dividend to complete the problem.

Examples:

$5\overline{\smash)1.3}$

$15\overline{\smash)2.7}$

$\begin{array}{r} .26 \\ 5\overline{\smash)1.30} \\ -1\ 0\downarrow \\ \hline 30 \\ -30 \\ \hline 0 \end{array}$

$\begin{array}{r} .18 \\ 15\overline{\smash)2.70} \\ -1\ 5\downarrow \\ \hline 120 \\ -120 \\ \hline 0 \end{array}$

S. $5\overline{\smash)1.7}$ S. $25\overline{\smash)1.5}$ 1. $2\overline{\smash).13}$ 2. $5\overline{\smash)3.1}$

3. $22\overline{\smash)45.1}$ 4. $24\overline{\smash)3.6}$ 5. $5\overline{\smash)0.2}$ 6. $95\overline{\smash)3.8}$

7. $20\overline{\smash)2.4}$ 8. $4\overline{\smash)6.3}$ 9. $5\overline{\smash)0.3}$ 10. $5\overline{\smash)2.09}$

1	
2	
3	
4	
5	
6	
7	
8	
9	
10	
Score	

Problem Solving

A girl was born in 1979. How old will she be in 2007?

Speed Drills	Review Exercises

+

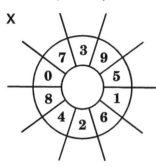

X

1. $2\overline{).15}$

2. $5\overline{).13}$

3. $100 \times 9.3 =$

4. $1,000 \times 9.3 =$

Move the decimal point in the divisor the number of places necessary to make it a whole number. Move the decimal point in the dividend the same number of places.

Examples:

$$.3\overline{)\begin{array}{r} .8 \\ 2.4 \\ -2\ 4 \\ \hline 0 \end{array}}$$

$$.03\overline{)\begin{array}{r} 950.\ * \\ 28.50. \\ -27\downarrow \\ \hline 15 \\ -15 \\ \hline 0 \end{array}}$$

*Sometimes placeholders are necessary.

S. $.7\overline{)2.73}$ S. $.15\overline{).036}$ 1. $.3\overline{)2.4}$ 2. $.03\overline{)5.1}$

3. $.9\overline{).378}$ 4. $.04\overline{)3.2}$ 5. $.06\overline{).324}$ 6. $2.1\overline{)6.72}$

7. $.26\overline{).962}$ 8. $.18\overline{).576}$ 9. $.04\overline{)2.3}$ 10. $.12\overline{)1.104}$

1	
2	
3	
4	
5	
6	
7	
8	
9	
10	
Score	

Problem Solving

If a car traveled 110.5 miles in two hours, what was its average speed per hour?

Speed Drills	**Review Exercises**

+

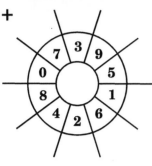

X

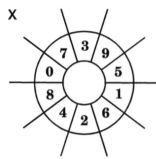

Helpful Hints

1. $3 \overline{)2.4}$ 2. $.03 \overline{)1.5}$

3. $.5 \overline{).19}$ 4. $.15 \overline{).6}$

To change fractions to decimals, divide the numerator by the denominator. Add as many zeroes as necessary.

Examples:

$\dfrac{3}{4}$ $4 \overline{)3.00} \quad .75$
$\underline{-\ 2\ 8\downarrow}$
20
$\underline{-\ 20}$
0

$\dfrac{3}{8}$ $8 \overline{)3.000} \quad .375$
$\underline{-\ 2\ 4\downarrow\downarrow}$
60
$\underline{-\ 56}$
40
$\underline{-\ 40}$
0

Change each of the following fractions to decimals.

S. $\dfrac{1}{2}$ S. $\dfrac{5}{8}$ 1. $\dfrac{3}{5}$ 2. $\dfrac{1}{4}$

3. $\dfrac{2}{5}$ 4. $\dfrac{7}{8}$ 5. $\dfrac{11}{20}$ 6. $\dfrac{13}{25}$

7. $\dfrac{5}{8}$ 8. $\dfrac{4}{20}$ 9. $\dfrac{1}{5}$ 10. $\dfrac{7}{10}$

1	
2	
3	
4	
5	
6	
7	
8	
9	
10	
Score	

Problem Solving

A worker earned 60 dollars and put $\frac{1}{4}$ of it in a savings account. How much did the worker put into the savings account?

Speed Drills	**Review Exercises**

+

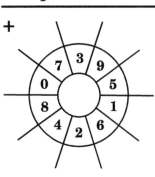

x

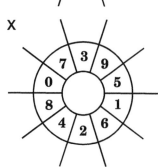

1. $3.36 + 5 + 2.6 =$

2. $3.2 - 1.63 =$

3.
$$3.12$$
$$\text{x} \quad .7$$

4. Change $\dfrac{7}{8}$ to a decmial

Helpful Hints	Use what you have learned to solve the following problems.	* Add as many zeroes as necessary. * Placeholders may be necessary. * Place decimal points properly.

S. $7\overline{)\,.035}$ S. $.06\overline{)\,2.4}$ 1. $3\overline{)\,2.28}$ 2. $5\overline{)\,.37}$

3. $1.6\overline{)\,.04}$ 4. $.3\overline{)\,1.35}$ 5. $.5\overline{)\,.12}$ 6. $.005\overline{)\,1.42}$

7. $.04\overline{)\,1.324}$ 8. $2.1\overline{)\,34.02}$ 9. Change $\dfrac{2}{5}$ to a decmial   10. Change $\dfrac{5}{8}$ to a decmial

1	
2	
3	
4	
5	
6	
7	
8	
9	
10	
Score	

Problem Solving	John weighed 120.5 pounds in January. By June he had lost 3.25 pounds. How much did he weigh in June?

Reviewing All Decimal Operations

1.
```
    3.72
    4.6
+   3.963
```

2. .3 + 2.96 + 7.1 =

3. 15.4 + 4 + 9.7 =

4.
```
    37.3
−   16.7
```

5.
```
    7.1
−   2.37
```

6. 6 − 1.43 =

7.
```
    3.12
x      3
```

8.
```
    3.4
x   16
```

9.
```
    .47
x   1.6
```

10.
```
    .436
x   3.21
```

11. 100 x 2.36 =

12. 1,000 x 2.7 =

13. $2 \overline{)2.68}$

14. $5 \overline{)7.3}$

15. $.5 \overline{).325}$

16. $.003 \overline{)1.2}$

17. $.15 \overline{).0075}$

18. $8.7 \overline{).1131}$

19. Change $\dfrac{7}{8}$ to a decmial

20. Change $\dfrac{11}{25}$ to a decmial

1	
2	
3	
4	
5	
6	
7	
8	
9	
10	
11	
12	
13	
14	
15	
16	
17	
18	
19	
20	

Speed Drills	**Review Exercises**

+

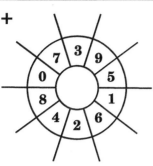

X

1. $\dfrac{3}{4} \div \dfrac{1}{2} =$ 2. $\dfrac{2}{3} \times 4\dfrac{1}{2} =$

3. $\dfrac{1}{2}$ 4. $\dfrac{2}{3}$
    $+\dfrac{2}{3}$ $-\dfrac{1}{5}$
    ___   ___

Helpful Hints	Percent means "per hundred" or "hundredths." If a fraction is expressed as hundredths, it can easily be written as a percent.	**Examples:** $\dfrac{7}{100} = 7\%$ $\dfrac{3}{10} = \dfrac{30}{100} = 30\%$ $\dfrac{19}{100} = 19\%$

Change each of the following to percents:

S. $\dfrac{17}{100} =$ S. $\dfrac{9}{10} =$ 1. $\dfrac{6}{100} =$ 2. $\dfrac{99}{100} =$

3. $\dfrac{3}{10} =$ 4. $\dfrac{64}{100} =$ 5. $\dfrac{67}{100} =$ 6. $\dfrac{1}{100} =$

7. $\dfrac{7}{10} =$ 8. $\dfrac{14}{100} =$ 9. $\dfrac{80}{100} =$ 10. $\dfrac{62}{100} =$

1	
2	
3	
4	
5	
6	
7	
8	
9	
10	
Score	

Problem Solving	A woman bought four chairs. If each chair weighed $22\dfrac{1}{2}$ pounds, what was the total weight of the chairs?

Speed Drills	**Review Exercises**

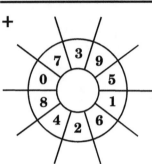

1. $\dfrac{7}{100} =$ _____ %

2. $\dfrac{9}{10} =$ _____ %

3. Find $\dfrac{1}{2}$ of $2\dfrac{1}{2}$

4. Find the difference between 17.6 and 9.85.

Helpful Hints	"Hundredths" = percent **Examples:** .27 = 27% .9 = .90 = 90%
	Decimals can easily be changed to percents *Move the decimal point twice to the right and add a percent symbol.

Change each of the following to percents:

S. .37	S. .7	1. .93	2. .02	**1**	
				2	
				3	
				4	
3. .2	4. .09	5. .6	6. .66	**5**	
				6	
				7	
7. .89	8. .6	9. .33	10. .8	**8**	
				9	
				10	
				Score	

Problem Solving	There are 32 fluid ounces in a quart. How many fluid ounces are there in .4 quarts?

Speed Drills	**Review Exercises**

+

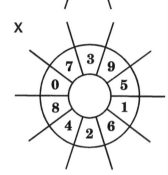

x

1. Reduce $\frac{18}{24}$ to its lowest terms.

2. Change $\frac{18}{16}$ to a mixed numeral with the fraction reduced to its lowest terms.

3. $5\frac{1}{5}$

                    $-\ 1\frac{1}{2}$
                    _____

4. $2\frac{1}{2}$

                    $+\ 3\frac{3}{5}$
                    _____

Helpful Hints	Percents can be expressed as decimals and as fractions. The fraction form may sometimes be reduced to its lowest terms.	**Examples:**	$25\% = .25 = \frac{25}{100} = \frac{1}{4}$ $8\% = .08 = \frac{8}{100} = \frac{2}{25}$

Change each percent to a decimal and to a fraction reduced to its lowest terms.

S. 20% = .   = ___   S. 9% = .   = ___   1. 16% = .   = ___

2. 6% = .   = ___   3. 75% = .   = ___   4. 40% = .   = ___

5. 1% = .   = ___   6. 45% = .   = ___   7. 12% = .   = ___

8. 5% = .   = ___   9. 50% = .   = ___   10. 13% = .   = ___

1	
2	
3	
4	
5	
6	
7	
8	
9	
10	
Score	

Problem Solving	75% of the students at Grover School take the bus. What fraction of the students take the bus? Reduce your fraction to its lowest terms.

Speed Drills	**Review Exercises**

+

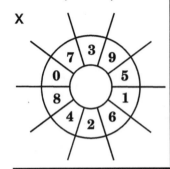

1. $.3\overline{).54}$

2. Change $\dfrac{4}{5}$ to a decimal.

X

3.   3.12
       x .6

4. $12 - 2.38 =$

Helpful Hints	To find the percent of a number, you may use either fractions or decimals. Use what is the most convenient.	**Example:** Find 25% of 60 .25 x 60	$\begin{array}{r} 60 \\ \times\ .25 \\ \hline 300 \\ 120 \\ \hline 15.00 \end{array}$	OR	$\dfrac{25}{100} = \dfrac{1}{4}$ $\dfrac{1}{\cancel{4}_1} \times \dfrac{\cancel{60}^{15}}{1} = \dfrac{15}{1} = 15$

S. Find 70% of 25.   S. Find 50% of 300.   1. Find 6% of 72.

2. Find 60% of 85.   3. Find 25% of 60.   4. Find 45% of 250.

5. Find 10% of 320.   6. Find 40% of 200.   7. Find 4% of 250.

8. Find 90% of 240.   9. Find 75% of 150.   10. Find 2% of 660.

1	
2	
3	
4	
5	
6	
7	
8	
9	
10	
Score	

Problem Solving	A gasoline tank holds $3\frac{3}{4}$ gallons. If $\frac{1}{3}$ of the tank has been used, then how many gallons have been used?

| **Speed Drills** | **Review Exercises** |

+

1. Find 13% of 85.　　2. Change $\frac{4}{5}$ to a decimal.

x

3. Change 3% to a decimal.　　4. Find 20% of 60.

Helpful Hints

When finding the percent of a number in a word problem, you can change the percent to a fraction or a decimal. Always express your answer in a short phrase or sentence.

Example:

A team played 60 games and won 75% of them. How many games did they win?

Find 75% of 60

.75 x 60

$$\begin{array}{r} 60 \\ \times .75 \\ \hline 300 \\ 420 \\ \hline 45.00 \end{array}$$

OR

$$\frac{75}{100} = \frac{3}{4}$$

$$\frac{3}{4_1} \times \frac{60^{15}}{1} = \frac{45}{1} = 45$$

Answer: The team won 45 games.

S. George took a test with 20 problems. If he got 15% of the problems correct, how many problems did he get correct?

S. If 6% of the 500 students enrolled in a school are absent, then how many students are absent?

1. A worker earned 80 dollars and put 70% of it into the bank. How many dollars did he put into the bank?

2. A car costs $9,000. If Mr. Smith has saved 20% of this amount, how much did he save?

3. Steve took a test with 30 problems. If he got 70% of the problems correct, how many problems did he get incorrect?

4. A family's monthly income is $3,000. If 25% of this amount is spent on food, how many dollars are spent on food?

5. There are 40 students in a class. If 60% of the class are boys, then how many girls are in the class?

6. A house that costs $80,000 requires a 20% down payment. How many dollars are required for the down payment?

7. If a car costs $6,000 and loses 30% of its value in one year, how much will the car be worth in one year?

8. A coat is priced $50. If the sales tax is 7% of the price, how much is the sales tax? What is the total cost including sales tax?

9. 23% of the 600 students at Madison School take instrumental music. How many students are taking instrumental music?

10. A family spends 25% of its income for food and 30% for housing. If its monthly income is $3,000, how much is spent each month on food and housing?

1	
2	
3	
4	
5	
6	
7	
8	
9	
10	
Score	

Problem Solving

A train traveled 83.5 miles per hour. At this rate, how far would it travel in 2.5 hours?

Speed Drills	**Review Exercises**

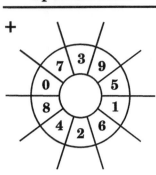

+

x

Helpful Hints

1. $20 \overline{)1764}$

2. $25 \times 36 =$

3. $9 + 19 + 216 + 3{,}674 =$

4. $7{,}010 - 6{,}914 =$

To change a fraction to a percent, first change the fraction to a decimal, then change the decimal to a percent. Move the decimal twice to the right and add a percent symbol.

Examples:

$\dfrac{3}{4}$ $4\overline{)3.00}$ $.75 = 75\%$
-28
20
-20
0

$\dfrac{16}{20} = \dfrac{4}{5}$ $5\overline{)4.00}$ $.80 = 80\%$
-40
0

* Sometimes the fraction can be reduced further.

Change each of the following to percents:

S. $\dfrac{1}{5} =$

S. $\dfrac{12}{15} =$

1. $\dfrac{3}{5} =$

2. $\dfrac{1}{2} =$

3. $\dfrac{1}{10} =$

4. $\dfrac{9}{12} =$

5. $\dfrac{15}{20} =$

6. $\dfrac{15}{25} =$

7. $\dfrac{1}{4} =$

8. $\dfrac{24}{30} =$

9. $\dfrac{18}{24} =$

10. $\dfrac{4}{20} =$

1	
2	
3	
4	
5	
6	
7	
8	
9	
10	
Score	

Problem Solving

320 people applied for jobs at a new department store. If 20% of the people were given a job, how many people got jobs?

Speed Drills	**Review Exercises**

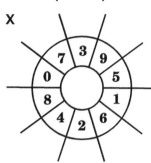

1.   4.19
      x 3

2.   12.6
    − 3.743

3.   36.16
      .724
    + 7.93

4.   .05 | .235

Helpful Hints

When finding the percent, first write a fraction, change the fraction to a decimal, then change the decimal to a percent.

Examples:

4 is what percent of 16?

$\frac{4}{16} = \frac{1}{4}$

    .25 = 25%
  4 | 1.00
    - 8↓
    ─────
     20
    - 20
    ─────
      0

5 is what percent of 25?

$\frac{5}{25} = \frac{1}{5}$

    .20 = 20%
  5 | 1.00
    - 1 0↓
    ─────
      00

S. 3 is what percent of 12?

S. 15 is what percent of 20?

1. 7 is what percent of 28?

2. 20 is what percent of 25?

3. 40 = what percent of 80?

4. 18 is what percent of 20?

5. 12 is what percent of 20?

6. 9 is what percent of 12?

7. 15 = what percent of 20?

8. 24 is what percent of 32?

9. 400 is what percent of 500?

10. 19 is what percent of 20?

1	
2	
3	
4	
5	
6	
7	
8	
9	
10	
Score	

Problem Solving

A man had 215 dollars in the bank. One day he withdrew 76 dollars and the next day he made a deposit of 96 dollars. How much does he now have in the bank?

Speed Drills	**Review Exercises**

+

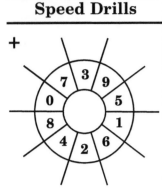

1.　Find 12% of 220.　　　2.　Change $\dfrac{3}{5}$ to a percent.

x

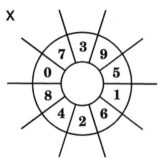

3.　Find 60% of 45.　　　4.　$.05\,\overline{\smash{)}1.7}$

Helpful Hints

When finding the percent first write a fraction, change the fraction to a decimal, then change the decimal to a percent.

Example:　　　　　15 is what % of 20?　　　$.75 = 75\%$　　They won 75% of
A team played 20 games　　　　　　　　　　　$4\,\overline{\smash{)}3.00}$　　the games.
and won 15 of them. What　　$\dfrac{15}{20} = \dfrac{3}{4}$　　　$\underline{-\ 28}$
percent of the games did　　　　　　　　　　　　20
they win?　　　　　　　　　　　　　　　　　　$\underline{-\ 20}$
　　　　　　　　　　　　　　　　　　　　　　　0

S.　A test had 20 questions. If Sam got 15 ques-tions correct, what percent did he get correct?

S.　In a class of 30 students, 12 are girls. What percent of the class is girls?

1.　On a spelling test with 25 words, Susan got 20 correct. What percent of the words did she get correct?

2.　A worker earned 200 dollars. If she put 150 dollars into a savings account, what percent of her earnings did she put into a savings account?

3.　A team played 16 games and won 12 of them. What percent did they lose?

4.　A quarterback threw 35 passes and 28 were caught. What percent of the passes were caught?

5.　$\dfrac{19}{20}$ of a class was present at school. What percent of the class was present?

6.　A class has an enrollment of 30 students. If 24 are present, what percent are absent?

7.　A team won 12 games and lost 13 games. What percent of the games played did it win?

8.　A school has 300 students. If 60 of them are sixth graders, what percent are sixth graders?

9.　On a math test with 50 questions Jill got 49 of them correct. What percent did she get correct?

10.　A pitcher threw 12 pitches. If 9 of them were strikes, what percent were strikes?

1	
2	
3	
4	
5	
6	
7	
8	
9	
10	
Score	

Problem Solving

There were 30 questions on a test. If a student got 80% of them correct, how many questions did he get correct?

Speed Drills	Review Exercises

+

x

Helpful Hints

1. Change $\frac{7}{100}$ to a percent.   2. Change $\frac{9}{10}$ to a percent.

3. Change .3 to a percent.   4. Change 24% to a decimal and a fraction expressed in its lowest terms.

Use what you have learned to solve the following problems.

Examples:

Find 15% of 35
.15 x 35

$$\begin{array}{r} 35 \\ \times\ .15 \\ \hline 175 \\ 35 \\ \hline 5.25 \end{array}$$

18 is what % of 24?

$$\frac{18}{24} = \frac{3}{4}$$

$$\begin{array}{r} .75 = 75\% \\ 4\ \overline{)\ 3.00} \\ -\ 28 \\ \hline 20 \\ -\ 20 \\ \hline 0 \end{array}$$

S. Find 20% of 45.

S. 3 is what percent of 12?

1. Find 3% of 120.

2. Find 80% of 72.

3. 12 is what percent of 16?

4. 20 = what percent of 25?

5. Find 25% of 310.

6. Find 12% of 50.

7. 12 is what percent of 48?

8. 4 = what percent of 40?

9. Find 90% of 500.

10. Find 22% of 236.

1	
2	
3	
4	
5	
6	
7	
8	
9	
10	
Score	

Problem Solving	A worker has completed $\frac{4}{5}$ of his project. What percent of his project has been completed?

Speed Drills

Review Exercises

+

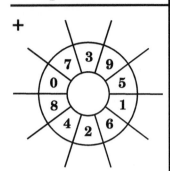

x

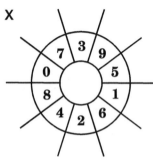

1. $2\frac{1}{2}$
$-2\frac{1}{3}$
_____

2. $3\frac{1}{4}$
$+2\frac{1}{2}$
_____

3. $\frac{7}{10} \div \frac{3}{14} =$

4. $\frac{4}{5} \div \frac{1}{3} =$

Helpful Hints

Use what you have learned to solve the following problems.

Examples:

A farmer has 210 cows. If he sells 40% of them, how many does he sell?

40% of 210   210
.4 x 210   x .4
                _____
                84.0

He sold 84 cows.

In a class of 24 students, 18 are girls. What percent are girls?

18 is what % of 24? $\frac{18}{24} = \frac{3}{4}$

75% are girls.

.75 = 75%

4) 3.00
    - 28
    ____
     20
    - 20
    ____
     0

S. A test has 40 problems. A student got 80% of them correct. How many problems did he get correct?

S. Sue has finished 6 problems on a test. If there are 24 problems on the test, what percent has she finished?

1. A ranch has 500 acres of land. If 60% of the land is used for grazing, then how many acres are used for grazing?

2. A player took 15 shots. If he made 9 of them, what percent did he make?

3. A man earned 24 dollars and spent 60% of it. How much did he spend?

4. A test has 45 questions. If Jane got 36 correct, what percent did she get correct?

5. 27 is what percent of 36?

6. Find 24% of 60.

7. There are 400 students in a school. If 60% eat cafeteria food, how many students eat cafeteria food?

8. A baseball team played 20 games and won 18. What percent did they lose?

9. A car costs $6,000. If a down payment of 20% is required, how much is the down payment?

10. 60 players tried out for a team. If only 12 made the team, what percent made the team? What percent didn't make the team?

1	
2	
3	
4	
5	
6	
7	
8	
9	
10	
Score	

Problem Solving

A student's test scores were 84, 96, 80, and 76. What was the student's average test score?

Change numbers 1 through 5 to a percent.

1. $\dfrac{17}{100}$ =   2. $\dfrac{3}{100}$ =   3. $\dfrac{7}{10}$ =   4. .19 =   5. .6 =

Change numbers 6 through 8 to a decimal and a fraction expressed in lowest terms.

6. 9% = . = _____   7. 14% = . = _____   8. 80% = . = _____

Solve the following problems. Label the word problem answers.

9. Find 4% of 320.   10. Find 60% of 230.

11. Find 12% of 600.   12. 3 is what percent of 5?

13. 12 is what percent of 15?   14. 12 is what percent of 48?

15. Change $\dfrac{1}{5}$ to a percent.   16. Change $\dfrac{1}{4}$ to a percent.

17. A man earned 200 dollars. If he put 40% of it into the bank, how many dollars did he put into the bank?

18. If a rancher has 450 cows and decides to sell 40% of them, how many cows will he sell?

19. A class has 40 students enrolled. If 18 are boys, what percent are boys?

20. A pitcher threw 80 pitches and 60 were strikes. What percent were strikes?

1	
2	
3	
4	
5	
6	
7	
8	
9	
10	
11	
12	
13	
14	
15	
16	
17	
18	
19	
20	

Speed Drills	Review Exercises

+

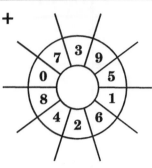

x

1. 3.6 + .72 + 3.9 =   2. 16.1 — 2.96 =

3.   1.64
          x .03

4.   5⟌3.0

Helpful Hints	Geometric term:	Point	Line	Plane	Line Segment	Ray
	Example:	• P	A B ⟷	plane ABC	A B	A B
	Symbol:	P	$\overleftrightarrow{AB}$	plane ABC	$\overline{AB}$	$\overrightarrow{AB}$

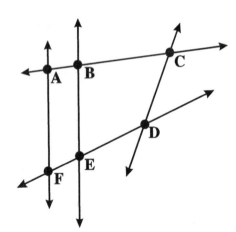

Use the figure to answer the following:

S. Name 4 points   S. Name 5 line segments

1. Name 5 lines   2. Name 5 rays

3. Name 3 points on   4. Give another name for
    $\overleftrightarrow{FD}$ $\overleftrightarrow{AB}$

5. Give another name for 6. Give another name for
    $\overleftrightarrow{ED}$ $\overrightarrow{AC}$

7. Name 2 line segments on $\overleftrightarrow{FD}$ 8. Name 2 rays on
                                   $\overleftrightarrow{FE}$

9. Name 2 rays on $\overleftrightarrow{AC}$ 10. What point is common to lines
                                 $\overleftrightarrow{FD}$ and $\overleftrightarrow{BE}$?

1	
2	
3	
4	
5	
6	
7	
8	
9	
10	
Score	

Problem Solving	If a factory can manufacture an engine in $2\frac{1}{2}$ hours, how long will it take to manufacture 10 engines?

Speed Drills	Review Exercises

+

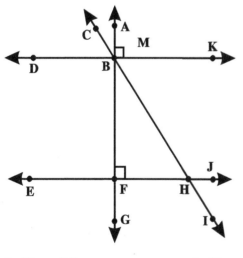

x

1.
$$\frac{3}{5}$$
$$+\ \frac{4}{5}$$
_____

2.
$$\frac{3}{4}$$
$$-\ \frac{1}{4}$$
_____

3. $2 \times \dfrac{3}{4} =$

4. $3 \div \dfrac{3}{4} =$

Helpful Hints

Geometric term: Example:	Parallel Lines	Intersecting Lines	Perpendicular Lines	Angle	Symbols
	⟷ ⟷	C — D / B — A	D / C—E / F	D / A Vertex C	∠DAC ∠CAD ∠A

Use the figure to answer the following:

S. Name 2 parallel lines

S. Name 2 perpendicular lines

1. Name 3 pairs of intersecting lines

2. Name 5 angles

3. Name 3 angles that have B as their vertex

4. Name 3 angles that have H as their vertex

5. Name 3 lines   6. Name 5 line segments   7. Name 5 rays

8. Name 3 line segments on $\overleftrightarrow{BH}$ 9. Name 3 lines which include point B

10. Give another name for ∠JHI

1	
2	
3	
4	
5	
6	
7	
8	
9	
10	
Score	

Problem Solving

Mary has a piece of cloth $7\frac{1}{2}$ yards long. How many pieces of $1\frac{1}{2}$ yards long can she cut from this piece?

Speed Drills	**Review Exercises**

+

x

1. 6 is what % of 8?

2. Find 15% of 225.

3. A man had 300 cows and decided to sell 15% of them. How many cows did he sell?

4. Sue took a test with 20 problems. If she got 14 of the problems correct, then what percent of the problems did she get correct?

Helpful Hints

right angle — measures 90°

acute angle — measures less than 90°

obtuse angle — measures more than 90°

straight angle — measures 180°

Use the figure to answer the following:

S. Name 4 right angles

S. Name 5 acute angles

1. Name 5 obtuse angles

2. Name 5 straight angles

3. What kind of angle is ∠IJG?

4. What kind of angle is ∠EDB?

5. What kind of angle is ∠GBD?

6. What kind of angle is ∠GJK?

7. Name an acute angle which has J as its vertex.

8. Name an obtuse angle which has D as its vertex.

9. Name a right angle which has B as its vertex.

10. Name a straight angle which has D as its vertex.

1	
2	
3	
4	
5	
6	
7	
8	
9	
10	
Score	

Problem Solving

A rope is 1.7 meters long. If a man wants to cut it into 5 pieces of equal length, how long will each piece be?

Speed Drills

+

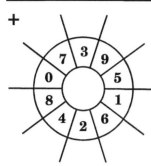

X

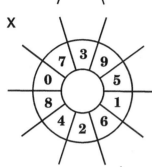

Review Exercises

1. Change $\frac{9}{7}$ to a mixed numeral

2. Change $7\frac{2}{3}$ to an improper fraction

3. Express $\frac{24}{40}$ in its lowest terms

4. $2\frac{1}{2} \times 3\frac{1}{2} =$

Helpful Hints

To use a protractor, following these rules:
1. Place the center point of the protractor on the vertex.
2. Place the zero mark on one edge of the angle.
3. Read the number where the other side of the angle crosses the protractor.
4. If the angle is acute, use the smaller number. If the angle is obtuse, use the larger number.

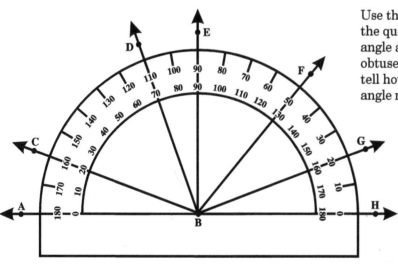

Use the figure to answer the questions. Classify the angle as right, acute, obtuse, or straight. Then tell how many degrees the angle measures.

1	
2	
3	
4	
5	
6	
7	
8	
9	
10	
Score	

S. ∠HBG   S. ∠DBH   1. ∠EBH   2. ∠CBH   3. ∠GBH   4. ∠DBA
acute   obtuse   right   obtuse   acute   acute

5. ∠ABF   6. ∠FBH   7. ∠ABH   8. ∠ABG   9. ∠EBA   10. ∠FBA
obtuse   acute   straight Obtuse   right   obtuse

Problem Solving

260 students took a social studies test and 80% received a passing grade. How many students received a passing grade?

Speed Drills	Review Exercises

+

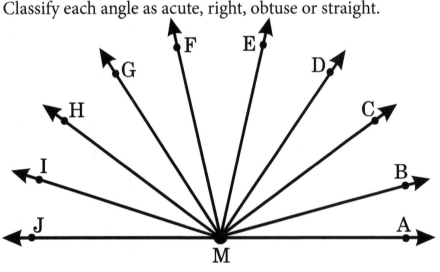

(Speed drill wheel: center 0, 7, 3, 9, 5, 8, 1, 4, 2, 6)

1. Change $\frac{15}{20}$ to a percent.   2.   Change .9 to a percent.

x

(Speed drill wheel: center 0, 7, 3, 9, 5, 8, 1, 4, 2, 6)

3. Find 4% of 65.

4. 12 is what percent of 30?

Helpful Hints

When using a protractor, remember to follow these tips:
1. Place the center point of the protractor on the vertex of the angle.
2. Place the zero edge on the edge of the angle.
3. Read the number where the other side of the angle crosses the protractor.
4. If the angle is acute, use the smaller number. If the angle is obtuse, use the larger number.

Classify each angle as acute, right, obtuse or straight.

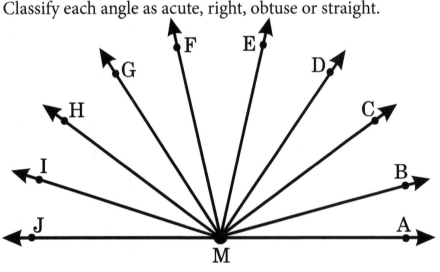

1	
2	
3	
4	
5	
6	
7	
8	
9	
10	
Score	

S. ∠AMC   S. ∠EMC   1. ∠DMA   2. ∠FMJ

3. ∠FMA   4. ∠DMJ   5. ∠EMA   6. ∠EMC

7. ∠FMB   8. ∠HMC   9. ∠IMA   10. ∠IMF

Problem Solving

There are 645 seats in an auditorium. If 379 of the seats are occupied, how many seats are empty?

Speed Drills	Review Exercises

+

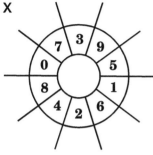

X

1. Name and classify this angle.

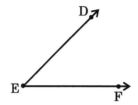

2. Name and classify this angle.

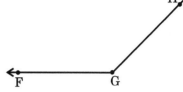

3. Name and classify this angle.

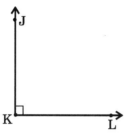

4. What kind of lines are these?

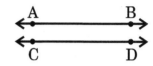

Helpful Hints — Polygons are closed figures made up of line segments.

triangle	rectangle	square	parallelogram	trapezoid
3 sides	4 sides, 4 right angles	4 congruent sides, 4 right angles	4 sides, opposite sides parallel	4 sides, 1 pair of parallel sides

Name each polygon. Some have more than one name.

S.

S.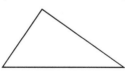

1.

2.

3.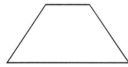

4.

5.

6.

7.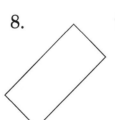

8.

9.

10.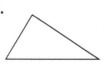

1	
2	
3	
4	
5	
6	
7	
8	
9	
10	
Score	

Problem Solving — A man earned $3.75 per hour. How much was his pay if he worked 8 hours?

Speed Drills

+

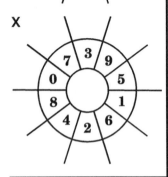

X

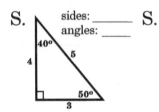

Review Exercises

1. $30 \overline{)7013}$

2.
```
    48
  x 36
```

3.
```
    732
     46
  + 377
```

4.
```
   7611
  -  799
```

	Sides			Angles		
Helpful Hints	Triangles can be classified by sides and angles.	equilateral △ 3 congruent sides	scalene △ no congruent sides	isosceles △ 2 congruent sides	acute △ 3 acute angles	right △ 1 right angle	obtuse △ 1 obtuse angle

Classify each triangle by its sides and angles.

S. sides: _____ angles: _____
40° 5 4 50° 3

S. sides: _____ angles: _____
60° 7 7 60° 60° 7

1. sides: _____ angles: _____
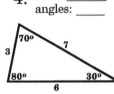
40° 7 12 120° 20° 9

2. sides: _____ angles: _____
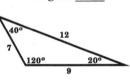
80° 7 7 50° 50° 9

3. sides: _____ angles: _____
45° 8 6 45° 6

4. sides: _____ angles: _____
70° 3 7 80° 30° 6

5. sides: _____ angles: _____
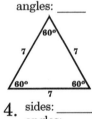

6. sides: _____ angles: _____
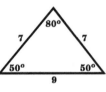
8 30° 5 40° 110° 3

7. sides: _____ angles: _____

60° 3 5 30° 4

8. sides: _____ angles: _____

9. sides: _____ angles: _____
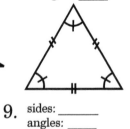
60° 9 ft. 9 ft. 60° 60° 9 ft.

10. sides: _____ angles: _____
45° 45°

1	
2	
3	
4	
5	
6	
7	
8	
9	
10	
Score	

Problem Solving

Buses hold 60 people. How many buses are needed for 143 people?

Geometry

Speed Drills

+

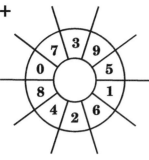

X

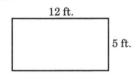

Review Exercises

1. Classify by sides.

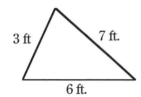

3 ft.   7 ft.
6 ft.

2. Classify by angles.

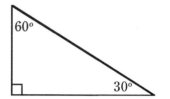

60°
30°

3. Classify by sides and angle.

sides: _____
angles: _____

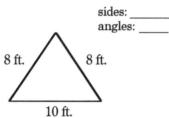

8 ft.   8 ft.
10 ft.

4. Find 60% of 75.

Helpful Hints

The distance around a polygon is its perimeter.

Examples:

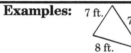

7 ft.
7 ft.
8 ft.

7
7
+ 8
perimeter = 22 ft.

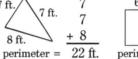

6 ft.

6
x 4
perimeter = 24 ft.

4 ft.
6 ft.

2 x (6 + 4) =
2 x (10) =
perimeter = 20 ft.

Find the perimeter of each of the following.

1	
2	
3	
4	
5	
6	
7	
8	
9	
10	
Score	

S.

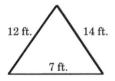

12 ft.
5 ft.

S.

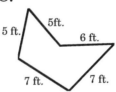

5ft.   5 ft.   6 ft.
7 ft.   7 ft.

1.

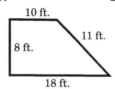

10 ft.
8 ft.   11 ft.
18 ft.

2.

12 ft.

3.

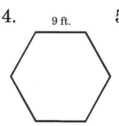

12 ft.   14 ft.
7 ft.

4.

9 ft.

5.

13 ft.
22 ft.

6.

10 ft.
8 ft.   8 ft.
15 ft.

7.

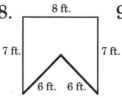

75 mi.   75 mi.
75 mi.

8.

8 ft.
7 ft.   7 ft.
6 ft.   6 ft.

9.

21 ft.
22 ft.

10.

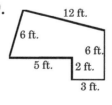

12 ft.
6 ft.
6 ft.
5 ft.   2 ft.
3 ft.

Problem Solving

A yard is in the shape of a rectangle which is 40 ft. wide and 55 ft. long. How many feet of fence will it take to go all the way around the yard?

| **Speed Drills** | **Review Exercises** |

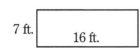

+

X

1. Find the perimeter.

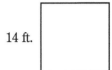

7 ft. | 16 ft.

2. Find the perimeter.

14 ft.

3. 6 is what % of 30?

4.
$$\frac{3}{4} - \frac{1}{3}$$

Helpful Hints — These are the parts of a circle.

chord, diameter, radius, center

* The length of the diameter is twice that of the radius.

Use the figure to answer the following:

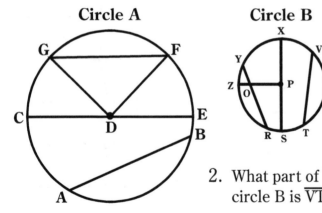

Circle A

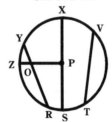

Circle B

S. What part of the circle is $\overline{CE}$?

S. Name 2 chords in circle B.

1. What part of circle A is $\overline{DF}$?

2. What part of circle B is $\overline{VT}$?

3. Name 3 radii in Circle A.

4. Name 2 chords in circle A.

5. If the length of $\overline{CE}$ is 16 ft., what is the length of $\overline{CD}$?

6. Name the center of Circle B.

7. Name 2 chords in circle B.

8. If $\overline{PS}$ in Circle B is 24 ft., what is the length of $\overline{XS}$?

9. Name 2 radii in Circle B.

10. Name a diameter in Circle B.

1	
2	
3	
4	
5	
6	
7	
8	
9	
10	
Score	

Problem Solving — A city is in the shape of a square. If its perimeter is 64 miles, what is the length of each side of the city?

Geometry

Speed Drills

Review Exercises

+

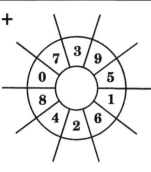

1. 325 + 16 + 9 =

2. 3 x 4.27 =

3. .3) 1.23

4. 7.6 + 14 + .3 + 2.14 =

X

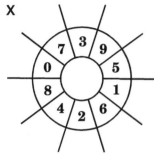

Helpful Hints

The distance around a circle is called its circumference. The Greek letter Π = pi = 3.14 or $\frac{22}{7}$. To find the circumference, multiply Π x diameter. Circumference = Π x d

Examples:

C = Π x d
= 3.14 x 6

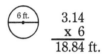

```
  3.14
x    6
18.84 ft.
```

$C = \Pi \times d$

$= \frac{22}{\cancel{7}_1} \times \frac{\cancel{14}^2}{1} = 44$ ft.

(Hint: If the diameter is divisible by 7, use Π = $\frac{22}{7}$)

Find the circumference of each of the following. If there is no figure, draw a sketch.

S.

4 ft.

S.

10 ft.

1.

6 ft.

2.

4 ft.

3. A circle with diameter 9 ft.

4. A circle with radius 14 ft.

5.

12 ft.

6.

5 ft.

7. A circle with radius 2 ft.

1	
2	
3	
4	
5	
6	
7	
Score	

Problem Solving

A garden is in the shape of a circle. If the radius is 12 feet, how far is it all the way around the garden?

Geometry
Areas of Squares and Rectangles

Speed Drills	Review Exercises

+

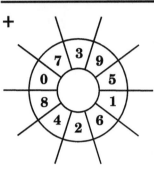

x

Helpful Hints

1.   3.2
    x 6.1

2. $\frac{3}{4}$ x $1\frac{1}{3}$ =

3. $2\frac{1}{2}$ x $1\frac{1}{5}$ =

4. 7 x 32.5 =

The number of square units needed to cover a region is called its area.

Examples:

area square = side x side

s = 7 ft.

A = s x s
A = 7 x 7   7
          x 7
          49 sq. ft.

area rectangle = length x width

w = 7 ft.   l = 12 ft.

A = l x w
A = 12 x 7   12
            x 7
            84 sq. ft.

Hint: 1. Start with formulas   2. Substitute values   3. Solve the problem

Find the following areas.

S. 13 ft.

S. 15 ft. 11 ft.

1. 14 ft. 6 ft.

2. 20 ft.

3. A rectangle with length 12 ft. and width 11 ft.

4. 2.5 ft. 4.3 ft.

5. $4\frac{1}{2}$ ft. $1\frac{1}{3}$ ft.

6. 25 ft.

7. A square with sides $2\frac{1}{2}$ ft.

1	84
2	100
3	132
4	10.75
⑤	
6	625
⑦	
Score	

Problem Solving	A floor is the shape of a rectangle. The length is 14 feet and the width is 13 feet. What is the area of the floor?

Geometry

Speed Drills

+

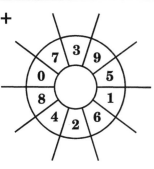

x

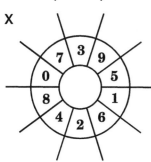

Helpful Hints

Review Exercises

1. Find the area

16 ft.

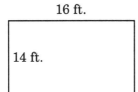

14 ft.

2. Find the area

16 ft.

3. Find the circumference

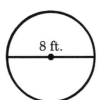

8 ft.

4. Find the circumference

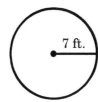

7 ft.

Area of a triangle = $\dfrac{\text{base x height}}{2} = \dfrac{b \times h}{2}$ Area parallelogram = base x height = b x h

Examples:

$A = \dfrac{b \times h}{2}$
height = 8 ft.
base = 7 ft.

$A = \dfrac{7 \times 8}{2} = \dfrac{56}{2}$

28 sq. ft.

2 ⟌ 56

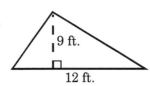
height = 5 ft.
base = 12 ft.

$A = b \times h$
$A = 12 \times 5$

$\begin{array}{r} 12 \\ \times\ 5 \\ \hline 60 \end{array}$ sq. ft.

Find the following areas.

S.
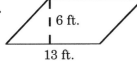
6 ft.
13 ft.

S.
11 ft.
14 ft.

1.
9 ft.
12 ft.

2.
11 ft.
16 ft.

3. A triangle with base 5 ft. and height 7 ft.

4. A parallelogram with base 13 ft. and height 7 ft.

5.

14 ft.
12 ft.

6.
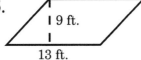
9 ft.
13 ft.

7.
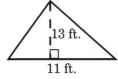
13 ft.
11 ft.

1	
2	
3	
4	
5	
6	
7	
Score	

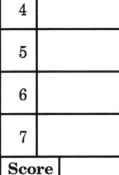

Problem Solving

A circular dodge-ball court has a diameter of 30 feet. How many feet is it around the dodge-ball court?

87

Speed Drills	**Review Exercises**

+

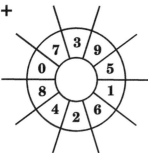

X

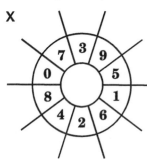

**Helpful
Hints**

1. Find the area

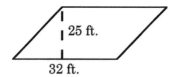

2. Find the area

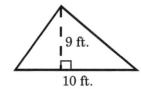

3. Find the area

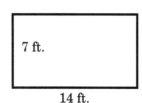

4. Find the area

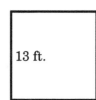

Area Circle = Π x radius x radius   A = Π x r x r

If the radius is divisible by 7, use $\Pi = \frac{22}{7}$

Examples: A = Π x r x r   3.14
                      = 3.14 x 3 x 3   x 9
                      = 3.14 x 9   28.26 sq. ft.

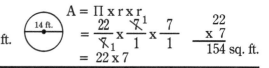

A = Π x r x r
$= \frac{22}{7_1} \times \frac{\cancel{7}^1}{1} \times \frac{7}{1}$ 22
                                        x 7
= 22 x 7   154 sq. ft.

Find the area of each circle.

S. 4 ft.

S. (12 ft.)

1. (5 ft.)

2. 14 ft.

3. 2 ft.

4. 8 ft.

5. 10 ft.

6. Circle with radius
   6 ft.

7. Circle with diameter
   14 ft.

1	
2	
3	
4	
5	
6	
7	
Score	

| **Problem
Solving**	A garden is in the shape of a circle. If the radius is 12 feet, how far is it all the way around the garden?

Geometry Reviewing Perimeter, Circumference, and Area

| **Speed Drills** | **Review Exercises** |

+

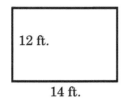

x

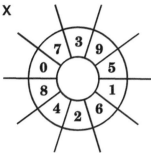

Helpful Hints

1. Find the perimeter

12 ft.

14 ft.

2. Find the area

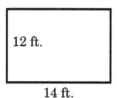

12 ft.

14 ft.

3. Find the circumference

6 ft.

4. Find the area

6 ft.

Remember these formulas.   For Areas: 1. Write formula 2. Substitute values 3. Solve problem

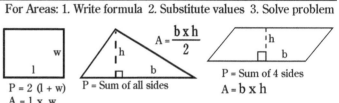

$C = \Pi \times d$ $P = 4 \times s$ $P = 2\,(l + w)$ P = Sum of all sides   P = Sum of 4 sides
$A = \Pi \times r \times r$ $A = s \times s$ $A = l \times w$ $A = \dfrac{b \times h}{2}$ $A = b \times h$

Find the perimeter or circumference. Second, find the area.

S.

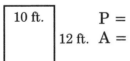

12 ft.

7 ft.

P =   A =

S.

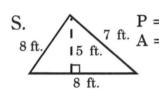

8 ft.   7 ft.   5 ft.   8 ft.

P =
A =

1.

12 ft.

P =
A =

2.

10 ft.

12 ft.

P =
A =

3.

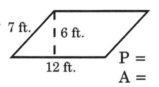

7 ft.   6 ft.   12 ft.

P =
A =

4.

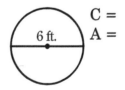

6 ft.

C =
A =

5.

14 ft.

C =
A =

6.

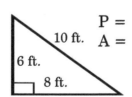

6 ft.   10 ft.   8 ft.

P =
A =

7. A square with sides 8 ft.

P =   A =

1	
2	
3	
4	
5	
6	
7	
Score	

Problem Solving

A man wants to buy a tent which is in the shape of a rectangle. If the length is 18 feet and the width is 12 feet, how many square feet of canvas will it take to make the tent?

Speed Drills	Review Exercises

+

x

1. $\quad \dfrac{3}{5}$
$\quad -\dfrac{1}{2}$
_____

2. $\quad \dfrac{2}{3}$
$\quad +\dfrac{1}{2}$
_____

3. $\quad 3\dfrac{1}{2} \text{ x } 3 =$

4. $\quad 2\dfrac{1}{2} \div \dfrac{1}{2} =$

Helpful Hints

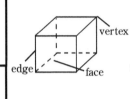 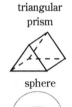 triangular prism, triangular pyramid, cone, cube, rectangular prism, sphere, square pyramid, cylinder

cones and cylinders do not have straight edges

Identify the shape and the numbers of each part.

1	
2	
3	
4	
5	
6	
7	
Score	

S.
name _____
faces _____
edges _____
vertices _____

S.
name _____
faces _____
edges _____
vertices _____

1.
name _____
faces _____
edges _____
vertices _____

2.
name _____
faces _____
edges _____
vertices _____

3.
name _____
faces _____
edges _____
vertices _____

4.
name _____
faces _____
edges _____
vertices _____

5.
name _____
faces _____
edges _____
vertices _____

6.
name _____

7. How many more faces does a cube have than a triangular prism?

Problem Solving

In a class of 40 people, 16 are girls. What percent of the class are girls? What percent are boys?

Geometry

Use the figure to answer questions 1 - 8

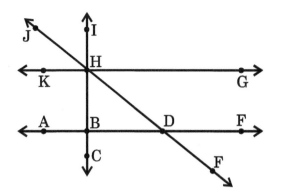

1. Name 2 parallel lines
2. Name 2 perpendicular lines
3. Name 4 line segments
4. Name 4 rays
5. Name 2 acute angles
6. Name 2 obtuse angles
7. Name 1 straight angle
8. Name 2 right angles

Triangle A

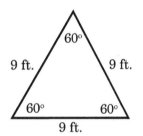

Triangle B

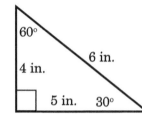

Use the figures to answer 9 and 10

9. Classify Triangle A by its sides and angles.

10. Classify Triangle B by its sides and angles.

11. Find the Perimeter

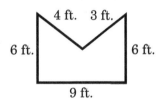

12. Find the Circumference

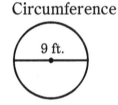

13. Find the Area

14. Find the Area

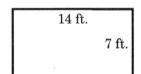

15. Find the Area

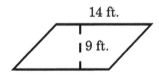

16. Find the Area

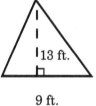

17. Find the Area

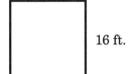

18. Identify and count the number of faces, edges, and vertices

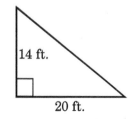

name _____
faces _____
edges _____
vertices _____

19. Find the Area

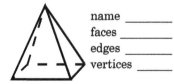

20. Find the perimeter of a square with sides of 96 feet.

1	
2	
3	
4	
5	
6	
7	
8	
9	
10	
11	
12	
13	
14	
15	
16	
17	
18	
19	
20	

Speed Drills	**Review Exercises**

+

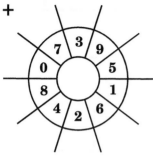

x

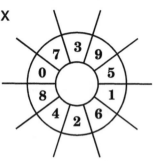

Helpful Hints

1. Find the area.

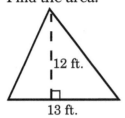

12 ft.

13 ft.

2. Find the circumference

13 ft.

3. 12 is what % of 16?

4. Find 3% of 425

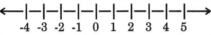

-4 -3 -2 -1 0 1 2 3 4 5

Integers to the left of zero are negative and less than zero. Integers to the right of zero are positive and greater than zero. When two integers are on a number line, the one farthest to the right is greater. **Hint:** Always find the sign of the answer first.

Examples:

The sum of two negatives is a negative.

-7 + -5 = -
(the sign is negative)

$$\begin{array}{r} 7 \\ + 5 \\ \hline 12 \end{array} = \text{(-12)}$$

When adding a negative and a positive, the sign is the same as the integer farthest from zero. Then subtract.

-7 + 9 = +
(the sign is positive)

$$\begin{array}{r} 9 \\ - 7 \\ \hline 2 \end{array} = \text{(+2)}$$

S. -9 + 12 =

2. -12 + -6 =

5. -8 + 32 =

8. 73 + -86 =

S. -15 + -6 =

3. 42 + -56 =

6. -39 + 76 =

9. -15 + -19 =

1. -15 + 29 =

4. -15 + -16 =

7. -96 + -72 =

10. 72 + -81 =

1	
2	
3	
4	
5	
6	
7	
8	
9	
10	
Score	

Problem Solving

560 students attend Lincoln School. If 40% of them ride the bus to school, then how many students don't ride the bus to school?

Speed Drills	Review Exercises

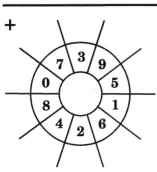

1. -16 + 18 =

2. 16 + -18 =

X

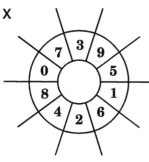

3. -16 + -18 =

4. Find the area.

3 ft.

Helpful Hints	When adding more than two integers, group the negatives and positives separately, then add.	**Examples:**

Examples:

-6 + 4 + -5 =   11
-11 + 4 = -   - 4
(sign is negative)   7 = -7

7 + -3 + -8 + 6 =   13
-11 + 13 = +   - 11
(sign is positive)   2 = +2

S. -3 + 5 + -6 =

S. -7 + 6 + -9 + 3 =

1. -3 + -4 + 5 =

2. 7 + -6 + -8 =

3. -15 + 19 + -12 =

4. -6 + 9 + 7 + 4 =

5. -16 + 32 + -18 =

6. -13 + 16 + -8 + 15 =

7. -9 + -7 + -6 =

8. -3 + 7 + -8 + -9 =

9. -32 + 16 + -17 + 8 =

10. -76 + 25 + -33 =

1	
2	
3	
4	
5	
6	
7	
8	
9	
10	
Score	

Problem Solving	The Giants played 36 games and won 27. What percent did they win? What percent did they lose?

Speed Drills	**Review Exercises**

+

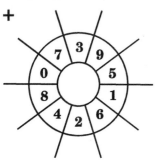

1. Find the perimeter of a rectangle with length 29 inches and width 13 inches.

2. -3 + 7 + -6 + 3 =

X

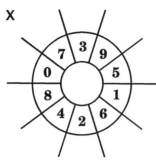

3. -12 + 27 + -6 =

4. -14 + -12 + 6 + 17 =

Helpful Hints	To subtract integers means to add the opposite.	**Examples:**

To subtract integers means to add the opposite.

Examples:

-3 - -8 =   8
-3 + 8 = +   - 3
(sign is positive) 5 = (+5)

8 - 10 =   10
8 + -10 = -   - 8
(sign is negative) 2 = (-2)

6 - -7 =   7
6 + 7 = +   + 6
(sign is positive) 13 = (+13)

S. -6 - 8 =

S. -6 - 9 =

1. 3 - -9 =

2. 15 - 18 =

3. -16 - -25 =

4. -16 - 12 =

5. 32 - -14 =

6. 35 - 14 =

7. -6 - 4 =

8. -64 - -53 =

9. -49 - 54 =

10. -63 - -78 =

1	
2	
3	
4	
5	
6	
7	
8	
9	
10	
Score	

Problem Solving	If eggs cost $1.29 per dozen, how much will 7 dozen cost?

| **Speed Drills** | **Review Exercises** |

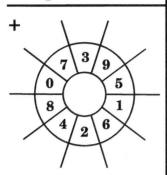

+

1. -6 - 9 =   2. -6 - -9 =

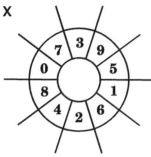

x

3. 16 - -18 =   4. -66 - 42 =

| **Helpful Hints** | Use what you've learned to solve the problems on this page. | **Examples:** |

$-7 + 4 + -3 + 2 =$ 10 $-7 - -6 =$ 7 $15 - 36 =$ 36
$-10 + 6 = -$ - 6 $-7 + 6 = -$ - 6 $15 + -36 = -$ - 15
(sign is negative) $\overline{4} = \textcircled{-4}$ (sign is negative) $\overline{1} = \textcircled{-1}$ (sign is negative) $\overline{21} = \textcircled{-21}$

S. -76 + 36 =   S. 9 - -6 =   1. -37 + -16 =

	1	

2. -92 + 103 =   3. -7 - 8 =   4. 6 - -9 =

	2	
	3	
	4	
	5	

5. -7 + 3 + -8 =   6. 14 + -6 + 3 + -8 =   7. 63 - 96 =

| | 6 | |
| | 7 | |

8. 3 - -12 =   9. -326 + 427 =   10. -273 - 408 =

	8	
	9	
	10	
	Score	

| **Problem Solving** | A business needs 325 postcards to mail to customers. If postcards come in packages of 25, how many packages does the business need to buy? |

Speed Drills	**Review Exercises**

+

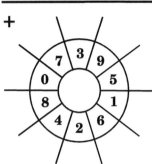

x

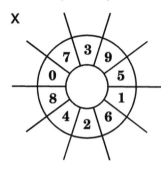

1. $3 \times 1\frac{1}{4} =$ 2. $6 \div 1\frac{1}{2} =$

3. $\frac{3}{4} \times 16 =$ 4. $\frac{4}{5} \div \frac{1}{10} =$

Helpful Hints	The product of two integers with different signs is negative. The product of two integers with the same sign is positive (• means multiply).	**Examples:** $7 \cdot -16 = -$ (sign is negative)	$\begin{array}{r} 16 \\ \times\ 7 \\ \hline 112 \end{array}$ = -112	$-8 \cdot -7 = +$ (sign is positive)	$\begin{array}{r} 8 \\ \times\ 7 \\ \hline 56 \end{array}$ = +56

S. $-3 \times -16 =$ S. $-18 \cdot 7 =$ 1. $-4 \cdot -17 =$

2. $16 \times -4 =$ 3. $-24 \cdot -12 =$ 4. $23 \times -16 =$

5. $-23 \cdot 32 =$ 6. $7 \times -19 =$ 7. $-3 \cdot -7 =$

8. $-19 \times -20 =$ 9. $32 \cdot -8 =$ 10. $-16 \cdot -12 =$

1	
2	
3	
4	
5	
6	
7	
8	
9	
10	
Score	

Problem Solving	At night the temperature was 37°. By morning it had dropped 47°. What was the temperature in the morning?

Speed Drills	**Review Exercises**

+

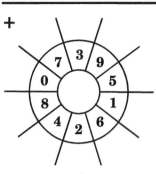

1. -6 + 7 + -2 + 6 =   2. 3 - -7 =

x

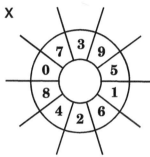

3. -3 • -9 =   4. -6 x -42 =

Helpful Hints	When multiplying more than two integers, group them in pairs to simplify. An integer next to a parentheses means to multiply. **Examples:**	2 • -3 (-6) = 6 (2 • -3) (-6) = x 6 -6 (-6) = + ‾36‾ =(+36) (sign is positive)	-2 • -3 • 4 • -2 = 8 (-2 • -3) • (4 • -2) = x 6 6 • -8 = - ‾48‾=(-48) (sign is negative)

S. -3 • 7 • -2 =   S. -3 (6) • -3 =   1. 2 (-3) • 4 =

	1	
	2	

2. -4 • -3 (-4) =   3. 2 • -3 • -4 • 5 =   4. 6 (3) • -4 x (-5) =

	3	
	4	
	5	

5. 1 • -1 • -3 • -2 =   6. (-2) (-3) (-4) =   7. -8 (-1) • 1 (-4) =

	6	
	7	
	8	

8. 4 (-3) • 2 (-3) =   9. (-3) (-2) (3) (4) =   10. 10 (-11) (-3) =

	9	
	10	
	Score	

Problem Solving	How much will a worker earn in 15 hours if he earns $1\frac{1}{2}$ dollars per hour?

Speed Drills	Review Exercises

+

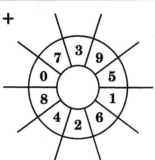

X

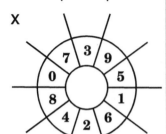

Review Exercises

1. $6\overline{)607}$

2. -3 (4) • -5 =

3. 12.3
   x 7

4. 7 + .63 + 7.18 =

Helpful Hints	The quotient of two integers with different signs is negative. The quotient of two integers with the same signs is positive. (HINT: Determine the sign, then divide.)	**Examples:**	$36 \div -4 = -$ (sign is negative)	$\begin{array}{r} 9 \\ 4\overline{)36} \\ -36 \\ \hline 0 \end{array}$ =-9	$\dfrac{-123}{-3} = +$ (sign is positive)	$\begin{array}{r} 41 \\ 3\overline{)123} \\ -12\downarrow \\ \hline 3 \end{array}$ =+41

S. 9 ÷ -3 =

S. $\dfrac{-90}{-15}$ =

1. -64 ÷ 4 =

2. -336 ÷ -7 =

3. $\dfrac{-75}{-5}$ =

4. 104 ÷ -4 =

5. $\dfrac{-110}{-5}$ =

6. 288 ÷ -12 =

7. 42 ÷ -7 =

8. 714 ÷ -21 =

9. $\dfrac{-65}{-5}$ =

10. 684 ÷ -36 =

1	
2	
3	
4	
5	
6	
7	
8	
9	
10	
Score	

Problem Solving	If the temperature at midnight was -7° and by 3:00 A.M. the temperature had dropped another 19°F, what was the temperature at 3:00 A.M.?

Speed Drills	Review Exercises

+

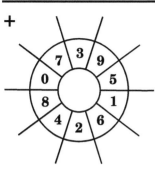

1. $-36 \div 4 =$

2. $\dfrac{-56}{-7} =$

X

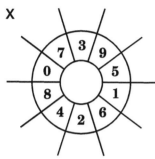

3. $3 \bullet 6 \bullet -5 =$

4. $-2 \, (-3) \, (-4) =$

Helpful Hints	Use what you have learned to solve problems like these.	**Examples:** $\dfrac{-36 \div -9}{4 \div -2} = \dfrac{4}{-2} = \boxed{-2}$ (sign is negative) $\qquad \dfrac{4 \times -8}{-8 \div 2} = \dfrac{-32}{-4} = \boxed{+8}$ (sign is positive)

S. $\dfrac{-10 \div 5}{2 \div -1} =$ S. $\dfrac{-4 \bullet -6}{-8 \div 4} =$ 1. $\dfrac{-32 \div -4}{-12 \div 3} =$

2. $\dfrac{-6 \times 5}{-30 \div 3} =$ 3. $\dfrac{12 \bullet -2}{18 \div 3} =$ 4. $\dfrac{-6 \bullet -6}{2 \div -2} =$

5. $\dfrac{54 \div -9}{-18 \div -9} =$ 6. $\dfrac{16 \div -2}{-1 \times -4} =$ 7. $\dfrac{-75 \div -25}{-3 \div -1} =$

8. $\dfrac{42 \div -2}{-3 \bullet -7} =$ 9. $\dfrac{45 \div -5}{-9 \div 3} =$ 10. $\dfrac{-56 \div -7}{-36 \div -9} =$

1	
2	
3	
4	
5	
6	
7	
8	
9	
10	
Score	

Problem Solving	A boy scored 627 points on a video game. This was 129 points more than his brother. How many points did his brother score?

Solve each of the following.

1. -9 + 7 =   2. 9 + -7 =   3. -9 + -7 =

4. -7 + -8 + 14 =   5. -32 + 16 + 21 + -24 =

6. 7 - 9 =   7. 4 - -9 =   8. -3 - 9 =

9. -13 - 14 =   10. 16 - 17 =   11. 3 • -16 =

12. -4 • -19 =   13. 2 (-7) (-4) =   14. -2 • 3 (-4) • 2 =

15. -36 ÷ 4 =   16. -126 ÷ -3 =   17. $\frac{-128}{-8}$ =

18. $\frac{-36 \div 2}{24 \div -4}$ =   19. $\frac{6 \cdot -3}{-54 \div -6}$ =   20. $\frac{-20 \cdot -3}{-30 \div -10}$ =

1	
2	
3	
4	
5	
6	
7	
8	
9	
10	
11	
12	
13	
14	
15	
16	
17	
18	
19	
20	

Charts and Graphs   Bar

Speed Drills	**Review Exercises**

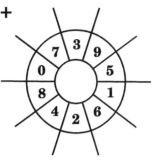

1. Find the area.

7 ft.

2. Find 15% of 65.

x

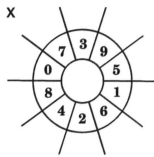

3. $\dfrac{3}{5} \times 2\dfrac{1}{2} =$

4. $3\dfrac{1}{2} \div 2 =$

Helpful Hints	Bar graphs are used to compare information.	1. Read the title. 2. Understand the meaning of the numbers. Estimate, if necessary. 3. Study the data. 4. Answer the questions, showing work if necessary.

Use the information in the graph to answer the questions.

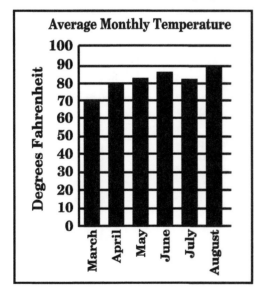

S. Which month had the second lowest average temperature?

S. How many degrees cooler was the average temperature in April than in August?

1. In which month was the average temperature 81°?

2. In which month did the average temperature drop from the previous month?

3. Which month had the second highest average temperature?

4. How much warmer was the average temperature in August than in May?

5. For what month did the average temperature rise the most from the previous month?

6. Which months had average temperatures less than July's average temperature?

7. Which two month's average temperatures were the closest?

8. The coolest day in August was 77°. How much less than the average temperature was this?

9. What was the increase in average temperature from May to June?

10. Which months had an average temperature less than May's?

1	
2	
3	
4	
5	
6	
7	
8	
9	
10	
Score	

Problem Solving	A fence around a garden is in the shape of a circle. If its diameter is 12 yards, what is the distance around the fence?

101

Charts and Graphs

Speed Drills	Review Exercises

+

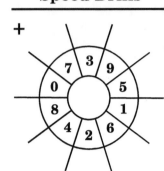

X

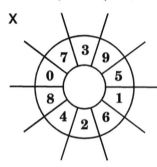

1. Change $\frac{15}{20}$ to percent.

2. Change .7 to a percent.

3. Find the area.

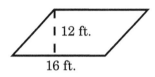

12 ft.

16 ft.

4. $12 \overline{)2643}$

Helpful Hints	1. Read the title. 2. Understand the meaning of the numbers. Estimate, if necessary. 3. Study the data. 4. Answer the questions, showing work if necessary.

Use the information in the graph to answer the questions.

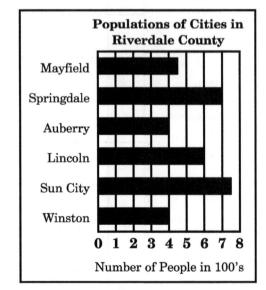

Populations of Cities in Riverdale County

Number of People in 100's

S. Which two cities had the same population?

S. What is the combined population of Mayfield and Lincoln?

1. Within 3 years, the population of Lincoln is expected to double. What will its population be in 3 years?

2. How many more people live in Springdale than in Auberry?

3. What is the difference in population between the largest and smallest cities?

4. What is the total population of Riverdale County?

5. How many more people live in Sun City than in Mayfield?

6. To reach a population of 900, by how much must Mayfield grow?

7. What is the total population of the two largest cities?

8. How many people must move to Winston before its population is equal to that of Springdale?

9. Which city has nearly double the population of Auberry?

10. What is the total population of all cities whose population is less than 500?

1	
2	
3	
4	
5	
6	
7	
8	
9	
10	
Score	

Problem Solving	A factory can make a part in $1\frac{1}{2}$ minutes. How many parts can be made in 30 minutes?

Charts and Graphs   Line

Speed Drills	**Review Exercises**

+

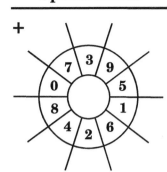

1. Find the area.

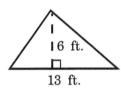

6 ft.

13 ft.

2. Classify the triangle.

sides _____

angles _____

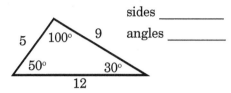

5   100°   9

50°   30°

12

X

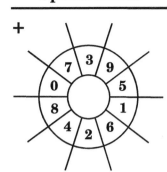

3. 1,000 x 2.365

4.   7.23

      x 6

Helpful Hints	Line graphs are used to show changes and relationships between quantities.	1. Read the title. 2. Understand the meaning of the numbers. Estimate, if necessary. 3. Study the data. 4. Answer the questions, showing work if necessary.

Use the information in the graph to answer the questions.

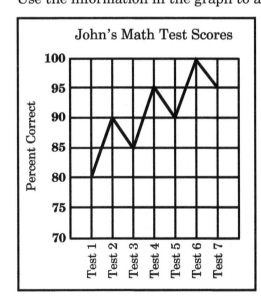

John's Math Test Scores

S. What was John's score on Test 5?

S. How much better was John's score on Test 7 than on Test 3?

1. On which three tests did John score the highest?

2. What is the difference between his highest and lowest score?

3. Find John's average score by adding all his scores and dividing by the number of scores.

4. How many scores were below John's average score?

5. What are his two lowest scores?

6. What is the average of his highest and lowest score?

7. How much higher was Test 4 than Test 1?

8. What was the difference between his highest score and his second lowest score?

9. How many test scores were improvements over the previous test?

10. Did John's progress generally improve or get worse?

1	
2	
3	
4	
5	
6	
7	
8	
9	
10	
Score	

Problem Solving	Movie passes cost $3.75 each. How much will it cost for 7 passes?

103

Speed Drills	Review Exercises

+

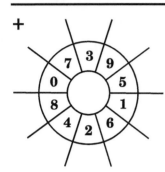

1. $\frac{4}{5}$
   $+ \frac{3}{5}$
   ———

2. $\frac{7}{8}$
   $- \frac{1}{8}$
   ———

X

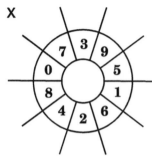

3. 3
   $- 1\frac{1}{7}$
   ———

4. $3\frac{1}{3}$
   $- 1\frac{2}{3}$
   ———

Helpful Hints

1. Read the title.
2. Understand the meaning of the numbers. Estimate, if necessary.
3. Study the data.
4. Answer the questions, showing work if necessary.

Use the information in the graph to answer the questions.

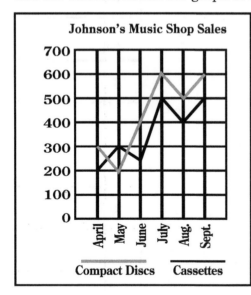

Johnson's Music Shop Sales

Compact Discs   Cassettes

S. In which months were the most cassettes sold?

S. How many more compact discs than cassettes were sold in August?

1. What was the total amount of compact discs sold in May and August?

2. How many more compact discs were sold in July than in August?

3. In which month were more cassettes sold than compact discs?

4. During which month did compact discs outsell cassettes by the most?

5. In September, how many more compact discs were sold than cassettes?

6. Which two months had the highest total sales?

7. What was the total number of compact discs and cassettes sold in September?

8. What was the increase in sales of compact discs between May and June?

9. What was the decrease in sales of compact discs between July and August?

10. What were the two lowest total sales months?

1	
2	
3	
4	
5	
6	
7	
8	
9	
10	
Score	

Problem Solving

Bill's test scores were 75, 80 and 100. What was his average score?

Charts and Graphs

Speed Drills	Review Exercises

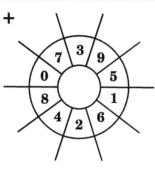

+

1. 32 + -76 =   2. 6 - -9 =

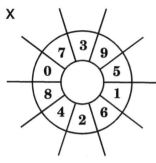

X

3. 3 • -7 =   4. Find 12% of 250.

Helpful Hints	A circle graph shows the relationship between the parts to the whole and to each other.	1. Read the title. 2. Understand the meaning of the numbers. Estimate, if necessary. 3. Study the data. 4. Answer the questions, showing work if necessary.

Use the information in the graph to answer the questions.

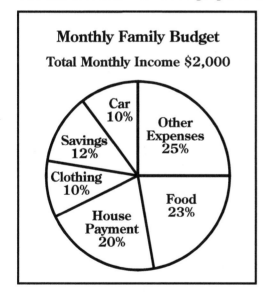

Monthly Family Budget

Total Monthly Income $2,000

Car 10%
Other Expenses 25%
Savings 12%
Clothing 10%
House Payment 20%
Food 23%

S. What percent of the family budget is spent for food?

S. After the car payment and house payment are paid, what percent of the budget is left?

1. What percent of the budget is used to pay for food and clothing?

2. How much is spent on food each month? (Hint: Find 23% of $2,000.)

3. How many dollars are spent on clothing?

4. How many dollars was the house payment?

5. What percent of the budget did the three largest items represent?

6. What percent of the budget is for savings, clothing, and the car payment?

7. In twelve months, what is the total income?

8. What percent of the budget is left after the house payment has been made?

9. Which two items require the same part of the budget?

10. Which part of the budget would pay for medical expenses?

1	
2	
3	
4	
5	
6	
7	
8	
9	
10	
Score	

Problem Solving	If a plane can travel 350 miles per hour, how far can it travel in 3.5 hours?

Speed Drills	Review Exercises

+

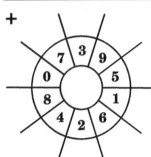

×

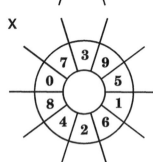

1. Reduce $\frac{12}{16}$ to its lowest terms.

2. Change $\frac{6}{8}$ to a percent.

3. 15 is what % of 20?

4. -33 + 16 =

Helpful Hints	Circle graphs can be used to show fractional parts.	1. Read the title. 2. Understand the meaning of the numbers. 3. Study the data. 4. Answer the questions.

Use the information in the graph to answer the questions.

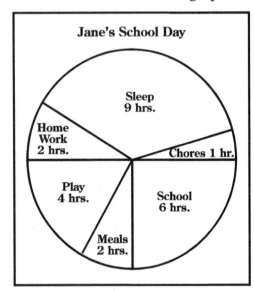

Jane's School Day

Sleep 9 hrs.

Home Work 2 hrs.

Chores 1 hr.

Play 4 hrs.

School 6 hrs.

Meals 2 hrs.

S. What fraction of the day does Jane play?

S. What fraction of Jane's day is used for school?

1. How many more hours does Jane sleep per day than play?

2. How many hours of homework does Jane have in a week?

3. How many hours per day are school- related activities?

4. What fraction of the day is spent for school, homework, and chores?

5. What fraction of the day does Jane spend for school, sleep, and chores?

6. How many hours does Jane spend in school in 3 weeks?

7. If Jane goes to bed at 9:00 P.M., what time does she get up in the morning?

8. If school starts at 8:30 A.M., what time is school dismissed?

9. How many hours per week does Jane spend in school and on homework?

10. What fractional part of the day does Jane play?

1	
2	
3	
4	
5	
6	
7	
8	
9	
10	
Score	

Problem Solving	A bedroom is in the shape of a rectangle with length 12 feet and width 14 feet. How many square feet of wall-to-wall carpet would it take to cover the floor?

Charts and Graphs

Speed Drills	Review Exercises

+

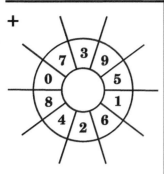

1. 7106 - 774 =

2. 76 x 403 =

x

3. 667 + 19 + 246 =

4. 5 | 5015

Helpful Hints	Picture graphs are another way to compare statistics.	1. Read the title. 2. Understand the meaning of the numbers. Estimate, if necessary. 3. Study the data. 4. Answer the questions.

Use the information in the graph to answer the questions.

Bikes Made By Street Bike Company

1986
1987
1988
1989
1990
1991

Each 🚲 represents 1,000 bikes

S. How many bikes were made in 1989?

S. How many more bikes were made in 1991 than in 1988?

1. Which year produced twice as many bikes as 1986?

2. What was the total number of bikes that the company produced in 1990 and 1991?

3. Which two years did the company make the most bikes?

4. 1992 is reported to be double the production of 1988. How many bikes are to be produced in 1992?

5. What is the total number of bikes produced in 1986 and 1991?

6. It cost $50 to make a bike in 1986, how much did the company spend making bikes that year?

7. The cost to make a bike jumped to $75 in 1991. How much did the company spend making bikes in 1991?

8. What is the difference in bikes produced in 1986 and 1991?

9. What is the total number of bikes made during the company's 3 most productive years?

10. One half of the bikes made in 1989 were ladies' style. How many ladies' bikes were made in 1989?

1	
2	
3	
4	
5	
6	
7	
8	
9	
10	
Score	

Problem Solving	Bill, John, Mary, and Sheila together earned $524. If they wanted to share the money equally, how much would each one receive?

Speed Drills	Review Exercises

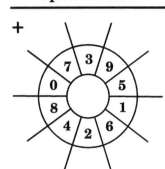

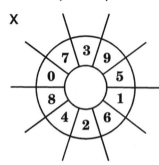

1. Find the perimeter of a rectangle with length 19 inches and width 13 inches.

2. Find the circumference of a circle with radius 3 feet.

3. Find the area of a rectangle with length 26 inches and width 15 inches.

4. Find the area of a triangle with base 8 feet and height 11 feet.

Helpful Hints

1. Read the title.
2. Understand the meaning of the symbols. Estimate, if necessary.
3. Study the data.
4. Answer the questions.

Use the information in the graph to answer the questions.

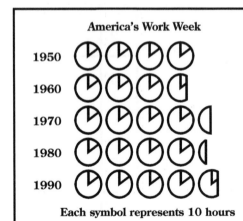

America's Work Week

1950
1960
1970
1980
1990

Each symbol represents 10 hours

S. Which work week was the longest?

S. How many hours shorter was the work week in 1960 than in 1990?

1. How many hours long was the workweek in 1970?

2. How may hours did the work week increase between 1980 and 1990?

3. If the average employee works 50 weeks per year how many hours did he work in 1950?

4. Which year's work week was approximately 38 hours?

5. Which years had the 2 shortest work weeks?

6. How many hours less was the work week in 1950 than 1980?

7. If the work week is 5 days, what was the average number of hours worked per day in 1950?

8. In 1990, if an employee decided to work 4 days per week, what would his average number of hours be per day?

9. Any work time over 40 hours is overtime. What was the average worker's weekly overtime in 1970?

10. What is the difference between the longest and the shortest work week?

1	
2	
3	
4	
5	
6	
7	
8	
9	
10	
Score	

Problem Solving

A man earned $5\frac{1}{4}$ dollars per hour. How much would he earn in 5 hours?

Reviewing Charts and Graphs

Height of Waterfalls

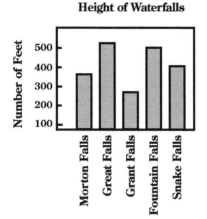

1. Which waterfall is the smallest?
2. Approximately how high is Great Falls?
3. Approximately how much higher is Morton Falls than Grant Falls?
4. Which waterfall is about the same height as Morton Falls?
5. Which waterfall is the fourth highest?

Family Budget: $3,000 per month

- Food 30%
- Housing 20%
- Clothing 13%
- Car 9%
- Savings 10%
- Misc. 18%

6. What percent of the family's money is spent on clothing?
7. What percent of the budget is spent on housing?
8. How many dollars are spent on housing per month? (Hint: Find 20% of $3,000)
9. How many dollars do they save per month?
10. What percent of the family budget is left after paying food, housing and car expenses?

Average Monthly Temperatures

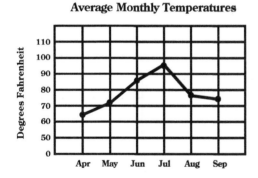

11. What is the average temperature of May?
12. How much cooler was April than July?
13. Which was the second hottest month?
14. Which month's temperature dropped the most from the previous month?
15. What is the difference in temperature between the hottest month and the second hottest month?

Fish caught in Drakes Bay in 1991

Salmon	🐟🐟🐟🐟🐟
Perch	🐟🐟🐟🐟
Cod	🐟🐟🐟🐟🐟
Bass	🐟🐟
Snapper	🐟🐟🐟🐟
Tuna	🐟🐟🐟

Each symbol represents 10,000 fish

16. How many perch were caught in 1991?
17. How many more snapper were caught than bass?
18. How many salmon and cod were caught?
19. If the average perch weighs 3 pounds, how many pounds were caught in 1991?
20. What are the three most commonly caught types of fish?

1	
2	
3	
4	
5	
6	
7	
8	
9	
10	
11	
12	
13	
14	
15	
16	
17	
18	
19	
20	

Speed Drills	Review Exercises

+

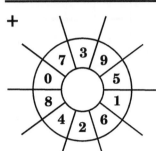

1. $376 + 92 + 743 =$ 2. $2,106 - 1,567 =$

x

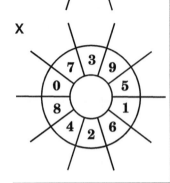

3.   724
         x 16

4.   7 ⟌ 1137

Helpful Hints	1. Read the problem carefully. 2. Find the important facts and numbers. 3. Decide what operations to use. 4. Solve the problem.	* Sometimes drawing a practice diagram can help. * Sometimes a formula is necessary. * Sometimes reading a problem more than once helps. * Show your work and put answers in a phrase.

S.   There are three sixth grade classes with enrollments of 36, 37, and 33. How many sixth graders are there in all?

S.   A family drove 355 miles each day for 7 days. How far did the family drive altogether?

1.   The attendance at the Hawk's game last year was 38,653. This year 45,629 attended. What was the increase in attendance this year?

2.   7 friends earned 1,463 dollars. If they want to divide the money evenly, how much will each person receive?

3.   A school put on a play. If 467 attended the Tuesday performance, 655 attended the Saturday performance, and 596 attended the Sunday performance, what was the total attendance?

4.   In an election, John received 2,637 votes. Julie received 2,904 votes. How many more votes did Julie receive than John?

5.   If a plane travels at 750 miles per hour, how far does it travel in 12 hours?

6.   A car traveled 295 miles in five hours. What was its average speed?

7.   How far can a car go if the tank contains 12 gallons of gas and it can travel 23 miles per gallon?

8.   A field is in the shape of a triangle with sides 265 feet, 379 feet and 189 feet. How many feet is it around the field?

9.   A rope is 544 feet long. If it is cut into pieces 2 feet long, how many pieces will there be?

10.   A library has 2,365 fiction books, 2,011 nonfiction books, and 796 reference books. How many books does the library have in all?

1	
2	
3	
4	
5	
6	
7	
8	
9	
10	
Score	

Problem Solving

Speed Drills	Review Exercises

+

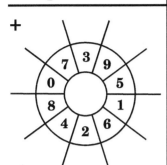

x

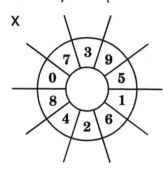

1. $\dfrac{3}{5}$

    $+\dfrac{1}{2}$
    _____

2. $\dfrac{7}{8}$

    $-\dfrac{1}{4}$
    _____

3. $2\dfrac{1}{2}$ x 3 =

4. $7\dfrac{1}{2} \div 1\dfrac{1}{2}$ =

Helpful Hints	1. Read the problem carefully. 2. Find the important facts and numbers. 3. Decide what operations to use. 4. Solve the problem.	* Sometimes drawing a practice diagram can help. * Sometimes a formula is necessary. * Sometimes reading a problem more than once helps. * Show your work.

S. A baker uses $2\frac{1}{4}$ cups of flour for a cake and $1\frac{3}{5}$ cups of flour for a pie. How much flour did he use?

S. A ribbon is $5\frac{1}{2}$ long. It was cut into pieces $\frac{1}{2}$ foot long. How many pieces were there?

1. Bill weighed $124\frac{1}{4}$ pounds three months ago. If he now weighs $132\frac{1}{2}$ pounds, how many pounds did he gain?

2. Steve earned 60 dollars and spent $\frac{2}{3}$ of it. How much did he spend?

3. It is $2\frac{1}{2}$ miles around a race track. How far will a car travel in 12 laps?

4. It takes a man $12\frac{1}{2}$ minutes to drive to work and $16\frac{3}{4}$ minutes to drive home. What is his total commute time?

5. What is the perimeter of a square flower bed which is $8\frac{1}{2}$ feet on each side?

6. Andy brought 6 melons, the total weight of which were 16 pounds. What was the average weight of each melon? (Express your answer as a mixed number.)

7. If a cook had $5\frac{1}{4}$ pounds of beef and used $3\frac{3}{4}$ pounds, how much beef was left?

8. Sue worked $7\frac{1}{2}$ hours on Monday and $6\frac{1}{4}$ hours on Tuesday. How many hours did she work in all?

9. A car can travel 50 miles per hour. At this rate, how far will it travel in $2\frac{1}{2}$ hours?

10. A factory can produce a tire in $2\frac{1}{2}$ minutes. How many tires can it produce in 40 minutes?

1	
2	
3	
4	
5	
6	
7	
8	
9	
10	
Score	

Speed Drills	Review Exercises

+

1. $3.26 + $4.19 + $6.24 =

2.
$$\begin{array}{r} \$7.00 \\ -\ \$3.56 \\ \hline \end{array}$$

x

3.
$$\begin{array}{r} \$3.92 \\ \times\ \ 6 \\ \hline \end{array}$$

4. $7\overline{\smash{)}\$15.33}$

Helpful Hints	1. Read the problem carefully. 2. Find the important facts and numbers. 3. Decide what operations to use 4. Solve the problem.	* Sometimes drawing a practice diagram can help. * Sometimes a formula is necessary. * Sometimes reading a problem more than once helps. * Show your work and put answers in a phrase.

S. If 6 dozen pencils cost $8.64, what is the cost of 1 dozen pencils?

S. Find the average speed of a jet that traveled 1350 miles in 2.5 hours.

1. Potatoes cost $3.19 a pound and carrots cost $2.78 per pound. How much more do potatoes cost per pound?

2. A man bought a chair for $125.50, a desk for $279.25, and a desk lamp for $24.95. What was the total cost of those items?

3. What is the area of a rectangular farm that is 1.8 miles long and 1.3 miles wide?

4. If 12 cans of corn cost $13.68, what is the cost of one can?

5. A plane can travel 240 miles in one hour. At this rate, how far can it travel in .8 hours?

6. A sack of potatoes cost $2.70. If the price was $.45 per pound, how many pounds of potatoes were in the sack?

7. A baseball mitt was on sale for $24.95. If the regular price was $35.50, how much could be saved by buying it on sale?

8. Tom weighs 129.6 pounds and Bill weighs 135.25 pounds. What is their total weight?

9. If 6 pounds of butter costs $5.34, what is the price per pound?

10. An engine uses 2.5 gallons of gas per hour. How many gallons will it use in 3.2 hours?

1	
2	
3	
4	
5	
6	
7	
8	
9	
10	
Score	

Speed Drills

+

X

Review Exercises

1. Find 16% of 230

2. What is 20% of 25?

3. Find the area

7 ft.

4. Find the area

16 ft.

9 ft.

Helpful Hints	Use what you have learned to solve the following problems.	1. Read the problem carefully. 2. Find the important facts and numbers. 3. Decide what operations to use 4. Solve the problem.

S. A rope 6 feet long is to be cut into pieces $1\frac{1}{2}$ feet long. How many pieces will there be?

S. If cans of soda cost $.35, how many cans of soda can be bought with $5.60?

1. What is the perimeter of a square field which has sides 139 feet long?

2. Steak cost $4.80 per pound. How much will .7 pounds cast?

3. Sam had a total score of 356 points on 4 tests. What was his average score?

4. A carpenter needs to glue together three boards with thickness of 3.9 inches, 2.25 inches, and 1.875 inches. What would the total thickness be after they are glued together?

5. A man earns $3\frac{1}{2}$ dollars per hour. How much will he earn in 8 hours?

6. A board was 9 feet long. If $2\frac{3}{4}$ feet was cut off, how much of the board was left?

7. After shopping for groceries a man had $13.64. He started with $50.00, how much did the groceries cost?

8. At a junior high school there are 600 seventh graders. If they are to be grouped into 15 equally sized homerooms, how many will be in each homeroom?

9. Sun City's population is 202,516. Elk Grove's population is 178,319. How much larger is Sun City's population?

10. A rancher owns 360 cows. If he decides to sell $\frac{3}{4}$ of them, how many will he sell?

1	
2	
3	
4	
5	
6	
7	
8	
9	
10	
Score	

Problem Solving

2-Step Whole Numbers

Speed Drills	Review Exercises

+

x

1. $2\overline{)73}$

2. $5\overline{)1172}$

3. $12\overline{)763}$

4. $25\overline{)5265}$

Helpful Hints
1. Read the problem carefully.
2. Find the important facts and numbers.
3. Decide what operations to use and in what order to do them.
4. Solve the problem.

S. Bill's test scores were 78, 87 and 96. What was his average score?

S. A store owner bought seven crates of lettuce that weighed 120 lbs. each. He also bought 12 sacks of potatoes that weighed a total of 125 lbs. What was the total weight of the lettuce and potatoes?

1. A man decides to buy a car. He makes a down payment of $500 and agrees to pay 36 monthly payment of $250. How much will the car cost him?

2. Last week a man worked 9 hours per day for five days. This week he worked 36 hours. How many hours did he work in all?

3. A farmer has an orchard with 36 rows of trees. There are 22 trees in each row. If each tree produces 14 bushels of fruit, how many bushels of fruit will be produced in all?

4. A junior high school has 7 eighth grade classes with 36 students in each class. If 506 students attend the school, how many are not in the eighth grade?

5. Buses hold 75 people. If 387 student and 138 parents are attending a football game, how many buses will they need?

6. A seventh grade has 314 boys and 310 girls. If they are to be grouped into equal classes of 26 each, how many classes will there be?

7. A car traveled 340 miles each day for 5 days and 250 miles on the sixth day. How many miles did it travel in all?

8. A tank holds 200,000 gallons of fuel. If 57,000 gallons were removed one day and 62,000 gallons the next, how many gallons are left?

9. Nine buses hold 85 people each. If 693 people buy tickets for a trip, how many seats won't be taken?

10. Jim worked 50 weeks last year. Each week he worked 36 hours. If he worked an additional 220 hours overtime, how many hours did he work in all last year?

1	
2	
3	
4	
5	
6	
7	
8	
9	
10	
Score	

114

It is illegal to photocopy this page. Copyright © 2018, Richard W. Fisher

Problem Solving   2-Step Fractions

Speed Drills	Review Exercises

+

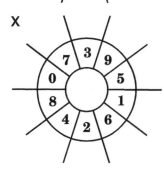

1. Find the average of 24, 42, and 45

2. Find $\frac{1}{2}$ of $3\frac{1}{2}$

x

3. $2\frac{6}{8} + 1\frac{1}{3} =$

4.
$$2\frac{3}{4}$$
$$+ 3\frac{1}{4}$$

Helpful Hints	1. Read the problem carefully. 2. Find the important facts and numbers. 3. Decide what operations to use. 4. Solve the problem.	* Sometimes drawing a practice diagram can help. * Sometimes a formula is necessary. * Sometimes reading a problem more than once helps. * Show your work and put answers in a phrase.

S. A tailor had $8\frac{1}{2}$ yards of cloth. He cut off 3 pieces which were $1\frac{1}{2}$ yards each. How much of the cloth was left?

S. A man had 56 dollars. If he gave $\frac{1}{2}$ of it to his son, how much did he have left?

1. A painter needs 7 gallons of paint. He already has $2\frac{1}{2}$ gallons in one bucket and $3\frac{1}{4}$ gallons in another. How many more gallons does he need?

2. There are 30 people in a class. If $\frac{2}{5}$ of them are boys then how many are girls?

3. Bill worked $5\frac{1}{2}$ hours on Monday and $6\frac{3}{4}$ hours on Tuesday. If he was paid 8 dollars per hour, what was his pay?

4. Susan made 36 bracelets last week and 42 this week. If she sold half of them, how many bracelets did she sell?

5. Joe's ranch has 4,000 acres. $\frac{1}{4}$ of the ranch was for crops. $\frac{2}{3}$ of the remainder was used for grazing. How many acres were for grazing?

6. A family was going to take a 400 mile trip. If they traveled $\frac{1}{4}$ of the distance the first day, how many miles were left to travel?

7. A store has 20 feet of copper wire. If the wire is cut into pieces $2\frac{1}{2}$ feet long and each piece sells for 6 dollars, how much would all the pieces cost?

8. A garden in the shape of a rectangle is 30 feet long and 20 feet wide. If $\frac{1}{3}$ of it is to be used for onions, how many square feet of the garden will be used for onions?

9. A farmer picked $6\frac{1}{2}$ bushels of fruit each day for 5 days. He then sold $15\frac{1}{2}$ bushels. How many bushels were left to sell?

10. A man bought $12\frac{1}{2}$ pounds of beef. He put $10\frac{1}{4}$ lbs. in his freezer. He used $\frac{3}{5}$ of the rest for cooking a stew. How many pounds did he use for the stew?

1	
2	
3	
4	
5	
6	
7	
8	
9	
10	
Score	

Speed Drills	**Review Exercises**

+

1. $4.3 + 7.6 + 7.97 =$

2. $\begin{array}{r} .63 \\ \times\ 3.5 \\ \hline \end{array}$

X

1. $2.013 - 1.66 =$

4. $3\overline{\smash{)}\$6.30}$

Helpful Hints	1. Read the problem carefully. 2. Find the important facts and numbers. 3. Decide what operations to use. 4. Solve the problem.	* Sometimes drawing a practice diagram can help. * Sometimes a formula is necessary. * Be careful about decimal placement. * Show your work and put answers in a phrase.

S. A man bought 5 bags of chips at $1.39 each, and a large pizza for $9.95. How much did he spend?

S. Mrs. Jones bought a hammer for $6.75 and a screwdriver for $5.19. If she gave the clerk a $20 bill, how much change would she receive?

1. A man bought a car. He paid $1500.00 down and $250 per month for 36 months. How much did he pay altogether?

2. Jill is taking a trip of 126 miles. If her car gets 21 miles per gallon and gas cost $1.19 per gallon, how much will the trip cost?

3. Susan worked 30 hours and was paid $4.15 per hour. If she bought a pair of shoes for $16.55, how much of her pay was left?

4. Cans of peas are on sale at 2 for $1.19. How much would 12 cans cost?

5. Jeans are on sale for $9.95 a pair. If the regular price is $12.25, how much would be saved by buying 3 pairs of jeans on sale?

6. Three friends earned $3.65 on Monday, $4.75 on Tuesday, and $3.75 on Wednesday. If they divided the money evenly, how much would each receive?

7. A jacket cost $12. If the sales tax was 7%, what was the total price of the jacket?

8. A man bought 12 gallons of gas at $1.16 per gallon and a quart of oil for $3.19. How much did he spend altogether?

9. A floor is 12 by 10 feet. If carpet costs $7.00 per square foot, how much will it cost to carpet the floor completely?

10. Cans of peas are 3 for $1.29. How much would one can of peas and a bag of chips priced at $2.19 cost altogether?

1	
2	
3	
4	
5	
6	
7	
8	
9	
10	
Score	

Problem Solving

Speed Drills

+

X

Review Exercises

1. Find 20% of 72

2. 6 is what percent of 24?

3. Find the circumference

6 ft.

4. Find the perimeter

26 inches

14 inches

Helpful Hints	Use what you have learned to solve the following problems.

S. A theater has 12 rows of 8 seats. If $\frac{2}{3}$ of them are taken, how many are taken?

S. Tom's times for the 100 yard dash were 11.8, 12.2 and 12.3 What was his average time?

1. John bought a shirt for $13.55 and a pair of shoes for $27.50. If he gave the clerk a $50 bill, how much change would he receive?

2. A group of hikers set out on a 50 mile trip. If they hiked 9 miles per day for three days, how many miles were left to hike?

3. Each of 3 classes has 32 students. If $\frac{2}{3}$ of these students are boys, how many boys are there?

4. If 2 pounds of chicken cost $3.60, how much will 10 pounds cost?

5. Mary is buying a bike. She makes a down payment of $75.00 and pays the rest in 12 monthly installments of $32.00 each. How much does she pay in all?

6. An electrician has 12 feet of wire. If he cuts it into pieces $1\frac{1}{2}$ feet long an sells each piece for $3.00, how many pieces are there and how much will they cost altogether?

7. Bill worked 35 hours at $5.00 per hour. If he wants to buy a bike for $225.00, how much more must he earn?

8. When a man started on a diet he weighed $160\frac{1}{2}$ pounds. The first week he lost $3\frac{1}{4}$ pounds and the second week he lost $2\frac{1}{2}$ pounds. How much did he weigh after 2 weeks?

9. A man bought 3 pens for $1.19 each, and a notebook for $3.15. How much did he spend altogether?

10. A car traveled 96 miles. If it averaged 12 miles per gallon of gas and gas cost $1.29, how much did the trip cost?

1	
2	
3	
4	
5	
6	
7	
8	
9	
10	
Score	

Speed Drills	Review Exercises

+

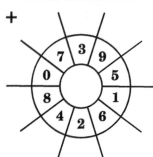

X

1. Classify by sides

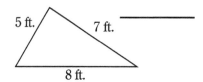

5 ft.   7 ft. _____

8 ft.

2. Classify by angles

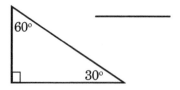

60° _____

30°

3. Find the perimeter of an equilateral triangle with sides of 39 feet.

4. Find the area of a square with sides of 16 feet.

Helpful Hints	1. Read the problem carefully. 2. Find the important number and facts. 3. Decide which operations to use and in what order 4. Solve the problem.	Re-read the problems. Draw a diagram or picture if necessary. Use a formula if necessary.

S. Bill bought 6 dozen hot dogs at $2.29 a dozen and 3 dozen burger patties at $6.15 a dozen. How much did he spend altogether?

1. Two trains leave a station in opposite directions, one at 85 miles per hour, the other at 75 miles per hour. How far apart will they be after 3 hours?

3. Mr. Jones' class has 32 students and Mrs. Jensen's class has 40 students. 25% of Mr. Jones' class got A's and 15% of Mrs. Jensen's class got A's. How many students got A's?

5. There are 400 boys and 350 girls at Anderson School. $\frac{3}{4}$ of the boys take the bus and $\frac{3}{5}$ of the girls take the bus. How many students in all take the bus?

7. Tom finished a 30 mile walkathon. He collected pledges of $26 for each of the first 25 miles and $36 for each remaining mile. How much did he collect in all?

9. A farm has 2000 acres. If $\frac{3}{5}$ of this was used for crops and $\frac{3}{4}$ of the remainder was used for grazing, how many acres were left?

S. A rectangular yard is 16 feet by 24 feet. How many feet of fencing is required to enclose it? If each 5-foot section costs $35.00, how much will the fencing cost?

2. A factory can make a table in 320 minutes and a chair in 8 minutes. How long will it take to make 7 tables and 15 chairs? Express the answer in hours and minutes.

4. John can buy a car by paying $3,000 down and making 36 monthly payment of $150 or by paying nothing down and making 48 payments of $180. How much would he save by paying the first way?

6. A farmer had 600 pounds of apples. He gave 200 pounds to neighbors and sold $\frac{1}{2}$ of the remainder at $.75 per pound. How much did he make selling the apples?

8. A developer has 3 plats of land. One was 39 acres, another was 37 acres, and another of 49 acres. How many $\frac{1}{4}$ acre lots did he have? If each lot sold for $2,000.00 how much did he sell all the lots for?

10. A carpenter bought 3 hammers for $7.99 each, 2 saws for $12.15 each and a drill for $26.55. How much did he spend in all?

1	
2	
3	
4	
5	
6	
7	
8	
9	
10	
Score	

Reviewing Problem Solving

1. Lincoln High School's enrollment is 5,679 and Jefferson High School's enrollment is 4,968. What is the total enrollment for both schools?

2. If a car travels 480 miles in 8 hours, what is its average speed?

3. A class has 35 students. If $\frac{2}{5}$ of them are boys, how many boys are in the class?

4. A rope $7\frac{1}{2}$ feet long was cut into $1\frac{1}{2}$ foot pieces. How many pieces were there?

5. Tim worked $12\frac{1}{4}$ hours on Monday and $9\frac{2}{3}$ hours on Tuesday. How much longer did he work on Monday?

6. A plane can travel 360 miles in one hour. How far can it go in .4 hours at this rate?

7. 6 pounds of corn costs $4.14. What is the price per pound?

8. A shirt was on sale at $12.19. If the regular price was $15.65, how much is saved by buying it on sale?

9. Bill's test scores were 78, 63 and 96. What was his average score?

10. A car traveled 265 miles each day for 7 days and 325 days on the eighth day. How many miles did it travel in all?

11. A tailor has 7 yards of cloth. He cut 2 pieces which were $1\frac{1}{3}$ yards each. How much cloth was left?

12. There are 45 students in a class. If $\frac{3}{5}$ of them are boys, how many are girls?

13. A car traveled 120 miles. It averaged 30 miles per gallon. If gas cost $1.09 per gallon, how much did the trip cost?

14. A boy bought 4 sodas for $.39 each and a bag of chips for $1.19. How much did he spend in all?

15. A rectangular living room floor is 12 feet by 14 feet. If carpet is $2.00 per square foot, how much will it cost to carpet the entire floor?

16. Cans of beans are 3 for $.69. How much would 12 cans cost?

17. A woman bought a car for $6,000.00. If she made a $2,000.00 down payment and paid the rest in equal payments of $400, how many payments would she make?

18. A ranch had 3,000 acres. $\frac{1}{3}$ of the ranch was used for grazing. $\frac{3}{5}$ of the rest was used for crops. How many acres were used for crops?

19. A plumber purchased 3 sinks for $129.00 each and 6 faucets for $16.35. How much did she spend in all?

20. A man wants to put a fence around a square lot with sides of 68 feet. How many feet of fencing is needed? If each 8-foot section of fence costs $25.00, how much will the fence cost?

1	
2	
3	
4	
5	
6	
7	
8	
9	
10	
11	
12	
13	
14	
15	
16	
17	
18	
19	
20	

Solve each of the following.

1.   347   2.   614   3.   6,403 + 763 + 16,799 =
       + 467   723
                            17
                        + 824

4.   6,502 + 2,134 + 654 + 24 =   5.   6,093 + 748 + 83 + 769 =

6.   927   7.   5,392   8.   6,053 − 4,639 =
       − 648   − 1,764

9.   5,000 − 3,286 =   10. 6,003 − 719 =   11.   73
                                                        x 4

12.   7,136   13.   45   14.   342
       x   4   x 37   x 46

15.   643   16. 4 ⌐ 526   17. 4 ⌐ 1376
       x 246

18. 40 ⌐ 568   19. 30 ⌐ 7614   20. 18 ⌐ 1243

1	
2	
3	
4	
5	
6	
7	
8	
9	
10	
11	
12	
13	
14	
15	
16	
17	
18	
19	
20	

Solve each of the following.

1. $\dfrac{4}{7}$
   $+ \dfrac{1}{7}$

2. $\dfrac{7}{8}$
   $+ \dfrac{3}{8}$

3. $\dfrac{3}{5}$
   $+ \dfrac{1}{3}$

4. $3\dfrac{1}{2}$
   $+ 2\dfrac{3}{8}$

5. $7\dfrac{3}{5}$
   $+ 6\dfrac{7}{10}$

6. $\dfrac{7}{8}$
   $- \dfrac{1}{8}$

7. $6\dfrac{1}{4}$
   $- 2\dfrac{3}{4}$

8. 5
   $- 2\dfrac{1}{7}$

9. $7\dfrac{3}{5}$
   $- 2\dfrac{1}{2}$

10. $7\dfrac{1}{4}$
    $- 2\dfrac{1}{3}$

11. $\dfrac{3}{5} \times \dfrac{2}{7} =$

12. $\dfrac{3}{20} \times \dfrac{5}{11} =$

13. $\dfrac{5}{6} \times 24 =$

14. $\dfrac{5}{8} \times 3\dfrac{1}{5} =$

15. $2\dfrac{1}{2} \times 3\dfrac{1}{2} =$

16. $\dfrac{5}{6} \div \dfrac{1}{3} =$

17. $2\dfrac{1}{3} \div \dfrac{1}{2} =$

18. $2\dfrac{2}{3} \div 2 =$

19. $5\dfrac{1}{2} \div 1\dfrac{1}{2} =$

20. $6 \div 1\dfrac{1}{3} =$

1	
2	
3	
4	
5	
6	
7	
8	
9	
10	
11	
12	
13	
14	
15	
16	
17	
18	
19	
20	

Solve each of the following.

1.   4.67
          3.5
      +   3.743

2.   .6 + 7.62 + 6.3 =

3.   16.8 + 6 + 7.9 =

4.   36.4
      −   17.8

5.   6.3
       −   3.69

6.   72 − 1.68 =

7.   2.64
       x   3

8.   2.6
        x 73

9.   .63
         x 2.4

10.   .126
       x 4.23

11.   10 x 3.65 =

12.   1,000 x 3.6 =

13. 2 $\overline{)\ 4.64\ }$

14. 5 $\overline{)\ 6.7\ }$

15. .4 $\overline{)\ .124\ }$

16. .004 $\overline{)\ 1.2\ }$

17. .15 $\overline{)\ .0045\ }$

18. .67 $\overline{)\ 8.71\ }$

19.   Change $\dfrac{3}{5}$ to a decimal.

20.   Change $\dfrac{7}{20}$ to a decimal.

1	
2	
3	
4	
5	
6	
7	
8	
9	
10	
11	
12	
13	
14	
15	
16	
17	
18	
19	
20	

Change numbers 1 through 5 to a percent.

1. $\dfrac{13}{100}$ =   2. $\dfrac{3}{100}$ =   3. $\dfrac{7}{10}$ =   4. .19 =   5. .6 =

Change numbers 6 through 8 to a decimal and a fraction expressed in lowest terms.

6. 8% = .   = ——   7. 18% = .   = ——   8. 80% = .   = ——

Solve the following problems. Label the word problem answers.

9. Find 3% of 74.   10. Find 40% of 320.

11. Find 16% of 400.   12. 4 is what percent of 5?

13. 18 is what % of 24?   14. 20 is what % of 25?

15. Change $\frac{16}{20}$ to a %.   16. Change $\frac{3}{5}$ to a percent.

17. 640 students attend Lincoln School. If 40% of the students are girls, how many girls attend Lincoln School?

18. A team played 40 games. If they won 65% of them, how many games did the team win?

19. A pitcher threw 40 pitches. If 30 were strikes, what percent were strikes?

20. Jim took a test with 20 questions. If he got 18 correct, what percent did he get incorrect?

	Percents
1	
2	
3	
4	
5	
6	
7	
8	
9	
10	
11	
12	
13	
14	
15	
16	
17	
18	
19	
20	

Final Review   Geometry

Use the figure to answer questions 1-8

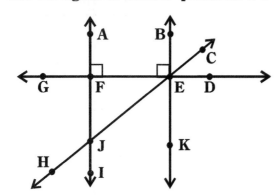

1. Name 2 parallel lines.
2. Name 2 perpendicular lines.
3. Name 4 line segments.
4. Name 4 rays.
5. Name 2 acute angles.
6. Name 2 obtuse angles.
7. Name 1 straight angle.
8. Name 2 right angles.

Triangle A   Triangle B

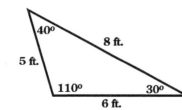

9. Classify triangle A by its sides and angles.
10. Classify triangle B by its sides and angles.

11. Find the perimeter.

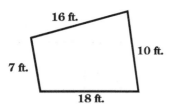

12. Find the circumference.

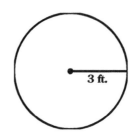

13. Find the area.

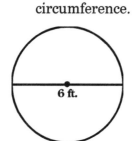

14. Find the area.

15. Find the area.

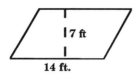

16. Find the area.

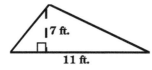

17. Find the area.

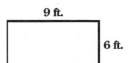

18. Identify and count the number of faces, edges, and vertices.

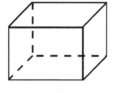

Name: Faces:
Edges   Vertices:

19. Find the area.

20. Find the perimeter of a square with sides of 37 ft.

1	
2	
3	
4	
5	
6	
7	
8	
9	
10	
11	
12	
13	
14	
15	
16	
17	
18	
19	
20	

1. -8 + 5 =

2. 8 + -5 =

3. -8 + -5 =

4. -6 + 3 + -4 + 2 =

5. -46 + 16 + 23 + -17 =

6. 6 - 9 =

7. 4 - -7 =

8. -3 - 9 =

9. -12 - 15 =

10. 15 - 19 =

11. 4 • -8 =

12. -3 • -19 =

13. 2 (-3) (-4) =

14. -2 • 3 • -6 =

15. -32 ÷ 8 =

16. -156 ÷ -3 =

17. $\dfrac{-136}{-8}$ =

18. $\dfrac{-32 \div -2}{16 \div 4}$ =

19. $\dfrac{5 \cdot -8}{-15 \div -3}$ =

20. $\dfrac{-40 \cdot -2}{-20 \div 2}$ =

1	
2	
3	
4	
5	
6	
7	
8	
9	
10	
11	
12	
13	
14	
15	
16	
17	
18	
19	
20	

Number of Aluminum Cans Collected at Allan School

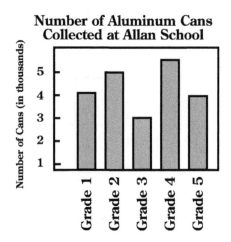

Fall Semester Science Grades

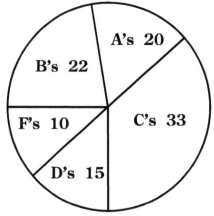

1. Which grade collected the most cans?

2. How many more cans did grade 4 collect than grade 1?

3. How many cans did grades 4 and 5 collect altogether?

4. How many cans were collected in all?

5. Which grade collected the second-most number of cans?

6. How many students got A's?

7. How many more C's than B's were there?

8. How many science students were there in all?

9. How many more C's and B's were there than A's and B's?

10. Which was the second largest group of grades?

Total Monthly Rainfall

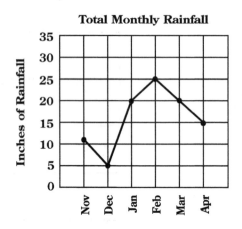

Girl Scout Cookies Sold During Fundraiser

Troop 15	● ● ● ● ◖
Troop 5	● ● ●
Troop 16	● ● ● ◖
Troop 11	● ● ● ● ●
Troop 3	● ● ●

Each symbol represents 100 boxes

11. How many inches of rain fell in February?

12. How many more inches of rain fell in January than in November?

13. Which month's rainfall increased the most from the previous month?

14. What was the total amount of rain for February, March and April?

15. Which month's rainfall decreased the most from the previous month?

16. How many boxes of cookies did Troop 11 sell?

17. Which troop sold the second most cookies?

18. How many more boxes did Troop 11 sell than Troop 5?

19. If Troop 3 sells three times as many boxes of cookies next year than they did this year, how much will they sell next year?

20. How many boxes of cookies did Troop 5 and Troop 15 sell together?

1	
2	
3	
4	
5	
6	
7	
8	
9	
10	
11	
12	
13	
14	
15	
16	
17	
18	
19	
20	

1. A man drove his car 396 miles on Monday and 476 miles on Tuesday. How many miles did he drive altogether?

2. A plane traveled 2,175 miles in 5 hours. What was its average speed per hour?

3. A student took a test with 32 problems. If he got $\frac{5}{8}$ of the problems correct, how many problems did he get correct?

4. 8 pounds of nuts were put into bags which each held $1\frac{1}{3}$ pounds. How many bags were there?

5. Steve earned $15\frac{1}{2}$ dollars on Monday and $12\frac{3}{5}$ dollars on Tuesday. How much more did he make on Monday?

6. A train can travel 215 miles in one hour. At this pace, how far can it travel in .7 hours?

7. 8 pounds of apples cost $12.48. How much does one pound cost?

8. Hats were on sale for $23.50. If the regular price was $30.25, how much would you save buying it on sale?

9. Bill weighed 90 pounds, Steve weighed 120 pounds and Mary weighed 84 pounds. What was their average weight?

10. Bill earned $45.00 each day for 5 days and $65.50 the sixth day. How much did he earn in all?

11. A highway crew intended to pave 8 miles of road. If they paved $2\frac{1}{3}$ miles the first week and $1\frac{1}{2}$ miles the second week, how much as left?

12. Bill took a test with 24 questions. If he missed $\frac{1}{6}$ of them, how many did he get correct?

13. A car traveled 174 miles and averaged 29 miles per gallon of gas. If gas was $1.15 per gallon, how much did the trip cost?

14. To tune up his car, Stan bought six spark plugs for $1.15 each and a condenser for $4.29. How much did he spend altogether?

15. A living room in a house was 14 by 8 feet and the dining room was 10 by 12 feet. How many square feet of wall-to-wall carpet would be needed to cover both floors?

16. Cans of corn are 3 for $.76. How much would 12 cans cost?

17. A man can pay for a car in 36 payments of 150 dollars or pay 4,800 dollars cash. How much could he save by paying cash?

18. Jean baked a pie. If she gave $\frac{1}{2}$ to Steve and $\frac{1}{3}$ to Gina, how much of the pie did she have left?

19. For a party, Mr. Jones bought six pizzas for $9.95 each and 3 cases of soda for $7.15 per case. How much did he spend in all?

20. A man has a rectangular lot which is 36 feet by 20 feet. How many feet of fence are needed to enclose it? If each 4-foot section costs 55 dollars, how much will the fence cost?

1	
2	
3	
4	
5	
6	
7	
8	
9	
10	
11	
12	
13	
14	
15	
16	
17	
18	
19	
20	

Page 5

1. 437
2. 735
3. 216
4. 12

S. 917
S. 1,484
1. 111
2. 1,102
3. 1,998
4. 1,186
5. 1,997
6. 675
7. 1,115
8. 207
9. 1,693
10. 105
Problem: 100 students

Page 6

1. 95
2. 823
3. 1,844
4. 95

S. 6,952
S. 29,994
1. 3,470,346
2. 51,422
3. 155,804
4. 1,879,958
5. 42,452
6. 83,749
7. 14,935
8. 31,116
9. 17,276
10. 82,646
Problem: 4,540,028

Page 7

1. 10,321
2. 40,719
3. 98,477
4. 94,257

S. 481
S. 2,461
1. 241
2. 399
3. 6,385
4. 5,877
5. 647
6. 3,193
7. 38
8. 3,070
9. 6,399
10. 339
Problem: 245 students

Page 8

1. 1,166
2. 574
3. 10,511
4. 5,649

S. 434
S. 221
1. 16
2. 435
3. 3,608
4. 264
5. 6,366
6. 1,468
7. 5,232
8. 62,420
9. 12,868
10. 1,999
Problem: $27,044

Page 9

1. 64
2. 6,429
3. 1,152,079
4. 4,533

S. 29,710
S. 3,655
1. 547
2. 573
3. 10,103
4. 7,243,008
5. 5,665
6. 1,196
7. 3,418
8. 1,845
9. 616
10. 90,558
Problem: 212 seats

Page 10

1. 4,385
2. 999
3. 572
4. 5,713

S. 1,269
S. 14,070
1. 201
2. 444
3. 2,292
4. 18,852
5. 16,288
6. 20,562
7. 56,406
8. 53,501
9. 58,476
10. 123,372
Problem: 2,190 days

Page 11

1. 2,052
2. 14,238
3. 5,439
4. 37,053

S. 1,058
S. 6,132
1. 4,526
2. 1,200
3. 1,692
4. 13,350
5. 7,110
6. 29,052
7. 448
8. 988
9. 67,528
10. 15,680
Problem: 832 desks

Page 12

1. 1,058
2. 14,472
3. 5,568
4. 997

S. 30,888
S. 337,666
1. 50,095
2. 167,564
3. 29,029
4. 99,138
5. 199,584
6. 78,900
7. 152,703
8. 130,662
9. 432,600
10. 155,644
Problem: 35,260 cars

Page 13

1. 2,142
2. 864
3. 10,844
4. 1,538

S. 7,584
S. 255,492
1. 162
2. 4,221
3. 32,508
4. 4,536
5. 14,202
6. 155,606
7. 5,676
8. 448,800
9. 9,020
10. 214,624
Problem: 12,000 sheets

Page 14

1. 249
2. 1,920
3. 138,863
4. 962

S. 5 r1
S. 9 r6
1. 8 r2
2. 5 r3
3. 12 r3
4. 15 r3
5. 12 r1
6. 7 r1
7. 9 r3
8. 24 r1
9. 12 r1
10. 18 r1
Problem: 12 boxes

Page 15

1. 6 r3
2. 15 r4
3. 14,938
4. 5,232

S. 144
S. 39 r3
1. 317 r1
2. 409 r1
3. 132
4. 136 r1
5. 145 r2
6. 158 r2
7. 121
8. 98 r4
9. 122 r3
10. 158 r5
Problem: 36 seats, 1 leftover

Page 16

1. 26 r1
2. 92 r5
3. 1,058
4. 152

S. 2,354
S. 863
1. 566
2. 3,371 r1
3. 1,122
4. 386 r3
5. 1,311
6. 1,498 r3
7. 10,453 r1
8. 11,980
9. 10,192 r1
10. 5,351
Problem: 144 boxes

Page 17

1. 203 r1
2. 7,416
3. 543
4. 22,059

S. 81 r2
S. 1,071
1. 50 r2
2. 1,500
3. 2,104 r2
4. 418 r5
5. 223 r2
6. 89 r4
7. 302 r5
8. 798 r5
9. 1,001 r1
10. 400 r7
Problem: 106 tickets

Page 18

1. 116 r6
2. 1,238
3. 271,800
4. 1,000 r3

S. 6 r7
S. 133 r22
1. 4 r62
2. 6 r39
3. 9 r13
4. 14 r27
5. 58 r23
6. 71 r3
7. 166 r24
8. 103 r11
9. 95 r1
10. 50 r26
Problem: 86 boxes

Page 19

1. 1,211
2. 156
3. 21,600
4. 1,167 r3

S. 21 r21
S. 31 r2
1. 22 r25
2. 21 r15
3. 10 r63
4. 7 r41
5. 21 r10
6. 31 r23
7. 11 r42
8. 11 r28
9. 45 r1
10. 21 r5
Problem: 13 classes

Page 20

1. 358
2. 17 r14
3. 21 r20
4. 20 r11

S. 3 r71
S. 19 r4
1. 5 r7
2. 6
3. 55 r7
4. 8 r10
5. 5 r3
6. 63 r4
7. 19 r3
8. 19 r31
9. 20 r1
10. 43 r13
Problem: 864 eggs

Page 21

1. 3 r6
2. 13 r22
3. 9 r36
4. 6 r19

S. 212 r22
S. 207 r15
1. 115
2. 234 r3
3. 102
4. 133 r10
5. 303 r10
6. 306
7. 73 r4
8. 40 r6
9. 45 r17
10. 41 r24
Problem: 125 pounds

Page 22

1. 24
2. 298 r4
3. 135 r36
4. 42 r17

S. 96 r40
S. 59 r9
1. 43 r1
2. 276 r4
3. 1,025 r2
4. 18 r7
5. 2 r56
6. 84 r19
7. 255 r9
8. 114 r11
9. 36 r26
10. 220 r9
Problem: 16 gallons

Page 23 Review

1. 1,011
2. 2,919
3. 80,653
4. 10,433
5. 8,447
6. 376
7. 3,915
8. 4,415
9. 2,322
10. 6,323
11. 228
12. 30,612
13. 2,438
14. 22.572
15. 232,858
16. 141 r2
17. 282 r5
18. 25 r19
19. 214 r4
20. 56 r9

Page 24

1. 434
2. 1,156
3. 1,387 r1
4. 1,296

S. 1/4
S. 5/6
1. 3/4, 6/8
2. 1/3
3. 2/4, 1/2
4. 3/8
5. 5/8
6. 2/3
7. 5/6
8. 1/8
9. 2/6, 1/3
10. 7/8
Problem: 384 crayons

Page 25

1. 2,142
2. 1,114
3. 494
4. 15 r17

S. 1/2
S. 3/4
1. 4/5
2. 3/4
3. 1/2
4. 4/5
5. 2/3
6. 2/3
7. 3/5
8. 5/8
9. 5/6
10. 3/4
Problem: 60 crayons

Page 26

1. 3/8
2. 2/3
3. 21 r7
4. 65,664

S. 1 ¾
S. 1 ½
1. 2 ½
2. 1 3/7
3. 2
4. 4 4/5
5. 4 1/2
6. 3
7. 4 1/2
8. 6 1/3
9. 1 1/2
10. 2 2/5
Problem: 2,828 more

Page 27

1. 1 2/7
2. 1 1/4
3. 5/7
4. 3/4

S. 3/5
S. 1 1/7
1. 3/4
2. 3/4
3. 3/5
4. 1/2
5. 1 1/4
6. 1 1/2
7. 1 2/3
8. 1 3/4
9. 1 3/4
10. 1 2/3
Problem: 2/3

Page 28
1. 3/5
2. 12 r9
3. 1 1/5
4. 368

S. 5 1/2
S. 6
1. 8
2. 7 1/2
3. 8
4. 8 1/6
5. 12 1/3
6. 6 1/5
7. 9 1/5
8. 12 1/2
9. 10
10. 8 1/7
Problem: 2 1/8 cups

Page 29
1. 5/6
2. 3 4/5
3. 1 1/2
4. 7 1/5

S. 1/2
S. 1/2
1. 1/2
2. 4/5
3. 3/4
4. 2/3
5. 3/4
6. 4/11
7. 1/2
8. 5/7
9. 5/12
10. 1/4
Problem: 3/5 miles

Page 30
1. 1/2
2. 1 1/5
3. 5 1/2
4. 943

S. 2 1/2
S. 2 2/3
1. 6 1/4
2. 4 1/2
3. 4 11/15
4. 6 2/3
5. 4 2/5
6. 3 2/5
7. 2
8. 2 3/5
9. 3 3/5
10. 2 3/5
Problem: 6 1/3 hours

Page 31
1. 1/2
2. 1 1/4
3. 8 1/3
4. 12 1/2

S. 3 2/5
S. 6 ¼
1. 3 3/7
2. 3 2/5
3. 6 1/3
4. 3 1/10
5. 4 7/8
6. 3 5/8
7. 3 2/9
8. 1/2
9. 3 7/10
10. 4 2/5
Problem: 2 1/8 yards

Page 32
1. 5/8
2. 1
3. 1 2/5
4. 4/5

S. 8 1/5
S. 4 2/3
1. 1 1/9
2. 3/4
3. 11 1/5
4. 4 2/3
5. 5 1/2
6. 4 2/3
7. 8 1/2
8. 4 2/3
9. 3 2/5
10. 2
Problem: 8 2/3 pounds

Page 33
1. 1 2/5
2. 1/2
3. 4 1/4
4. 3 3/5

S. 12
S. 24
1. 15
2. 18
3. 14
4. 24
5. 36
6. 20
7. 39
8. 60
9. 48
10. 36
Problem: 750 mph

Page 34
1. 200 r2
2. 1,548
3. 352
4. 461

S. 7/12
S. 1/2
1. 5/9
2. 1/6
3. 1 1/6
4. 11/15
5. 5/12
6. 1 1/14
7. 1 1/2
8. 17/22
9. 1/14
10. 1 5/36
Problem: 3 5/8 gallons

Page 35
1. 188 r2
2. 3/4
3. 1 1/10
4. 5/12

S. 7 11/12
S. 8 1/10
1. 8 1/6
2. 10 3/4
3. 8 1/8
4. 8 1/2
5. 15 3/4
6. 9 13/20
7. 5 3/10
8. 11 1/18
9. 8 8/15
10. 11 11/12
Problem: 4,224 parts

Page 36

1. 13/14
2. 10
3. 4 2/5
4. 1 2/3

S. 3 1/20
S. 2 5/6
1. 2 5/8
2. 7 1/2
3. 2 7/12
4. 4 5/8
5. 13/14
6. 5 13/16
7. 3 2/9
8. 3 5/6
9. 4 13/20
10. 2 3/8
Problem: 24 students

Page 37

1. 4/5
2. 7 1/4
3. 12
4. 1 1/30

S. 1 3/7
S. 10 ¼
1. 7/18
2. 13/18
3. 4 2/5
4. 8 5/8
5. 11 1/4
6. 4 3/4
7. 8 7/15
8. 3 1/2
9. 11/16
10. 8 7/12
Problem: 3 3/4

Page 38

1. 121
2. 26,100
3. 221
4. 6,268

S. 15/28
S. 12/25
1. 2/63
2. 1/5
3. 2 1/10
4. 14/27
5. 2/3
6. 1 1/15
7. 1 1/5
8. 6/35
9. 2/5
10. 1 1/14
Problem: 13/20 pounds

Page 39

1. 4/7
2. 1 13/15
3. 3/4
4. 1 1/3

S. 3/7
S. 1 1/2
1. 3/8
2. 1/10
3. 7/18
4. 3 1/3
5. 3/10
6. 2/3
7. 2/5
8. 9/20
9. 1 1/7
10. 10/21
Problem: 390 seats

Page 40

1. 3/7
2. 14/27
3. 13/15
4. 1/12

S. 9
S. 3 1/3
1. 6
2. 4
3. 20
4. 1 1/7
5. 13 1/2
6. 2 1/2
7. 3 1/2
8. 7 1/2
9. 15
10. 5 3/5
Problem: 24 girls

Page 41

1. 12 r40
2. 12
3. 6 2/3
4. 7/2

S. 3/4
S. 3
1. 1 7/8
2. 3 1/9
3. 16
4. 3
5. 6
6. 3 3/8
7. 8 1/8
8. 15
9. 11 2/3
10. 1 3/10
Problem: 14 miles

Page 42

1. 12/35
2. 27/100
3. 18
4. 8 2/3

S. 3/10
S. 7 1/2
1. 7/10
2. 1/6
3. 21
4. 5 1/7
5. 1 7/8
6. 1 5/6
7. 17
8. 2 1/3
9. 15 1/5
10. 8 1/8
Problem: 22 1/2 tons

Page 43

1. 14/15
2. 1/4
3. 7
4. 1 7/9

S. 1 1/3
S. 3/7
1. 1/6
2. 1 1/7
3. 4/13
4. 1/13
5. 2 1/2
6. 7
7. 1/9
8. 4 1/2
9. 2/9
10. 1/15
Problem: 45 dollars

Page 44

1. 1/9
2. 3 1/2
3. 3/11
4. 3 1/5
S. 1 1/7
S. 1 3/4
1. 2 1/4
2. 1 1/2
3. 7 1/2
4. 9
5. 4 2/3
6. 9
7. 1/2
8. 2 3/4
9. 3
10. 1 7/15
Problem: 7 pieces

Page 45 Review

1. 4/5
2. 1 1/3
3. 13/15
4. 8 2/9
5. 10 1/8
6. 1/2
7. 4 4/5
8. 4 2/5
9. 6 1/4
10. 5 14/15
11. 8/21
12. 3/26
13. 27
14. 1 7/8
15. 8 1/6
16. 1 1/2
17. 7
18. 2 4/9
19. 3 1/3
20. 2 4/7

Page 46

1. 1,876
2. 1 4/15
3. 428
4. 7/12
S. Two and six tenths
S. Thirteen and sixteen thousandths
1. seventy-three hundredths
2. four and two thousandths
3. one hundred thirty-two and six tenths
4. one hundred thirty-two and six hundredths
5. seventy-two and six thousand three hundred ninety-five ten thousandths
6. seventy-seven thousandths
7. nine and eighty-nine hundredths
8. six and three thousandths
9. seventy-two hundredths
10. one and six hundred sixty-six thousandths
Problem: 21 boys

Page 47

1. 8,424
2. 6/11
3. 5
4. 1 2/3
S. 6.04
S. 306.15
1. 9.8
2. 46.013
3. .0326
4. 50.039
5. .00008
6. .000004
7. 12.0036
8. 16.024
9. 23.5
10. 2.017
Problem: 3 2/5 degrees

Page 48

1. 1 1/2
2. 3/4
3. 2
4. 8 1/3
S. 7.7
S. 9.007
1. 16.32
2. .0097
3. 72.09
4. 134.0092
5. .016
6. 44.00432
7. 3.096
8. 4.901
9. 3.000901
10. .01763
Problem: 36 inches

Answer Key

Page 49

1. 13/15
2. 5 2/3
3. 1
4. 4/33

S. 1 43/100
S. 7 6/1,000
1. 173 16/1,000
2. 16/100,000
3. 7 14/1,000,000
4. 19 936/1,000
5. 9163/100,000
6. 77 8/10
7. 13 19/1,000
8. 72 9/10,000
9. 99/100,000
10. 63 143/1,000,000
Problem: 120 seats

Page 50

1. 5/6
2. 4 3/8
3. One and nineteen thousandths
4. 72 8/1,000

S. <
S. >
1. <
2. >
3. >
4. >
5. >
6. <
7. <
8. >
9. <
10. <
Problem: 8 miles

Page 51

1. 1 3/4
2. 1 1/4
3. 1
4. 14 2/5

S. 18.82
S. 8.84
1. 41.353
2. 14.431
3. 29.673
4. 1.7
5. 19.26
6. 146.983
7. 1.7
8. 25.044
9. 42.055
10. 31.09
Problem: 15.8 inches

Page 52

1. 13.726
2. 15.76
3. 1 3/4
4. 72.009

S. 13.84
S. 7.48
1. 5.894
2. 2.293
3. 1.16
4. 11.13
5. .053
6. 3.9577
7. 2.373
8. 3.684
9. 8.628
10. 26.765
Problem: 1.3 seconds

Page 53

1. 2/3
2. 3/4
3. 1 1/4
4. 2 1/5

S. 18.38
S. 3.687
1. 23.714
2. 6.17
3. 16.57
4. 8.24
5. 2.08
6. 19.56
7. 2.2
8. 4.487
9. 10.396
10. 25.7
Problem: $48.72

Page 54

1. 216
2. 1,472
3. 4,807
4. 265,608

S. 7.38
S. 36.8
1. 1.929
2. 14.64
3. 6.88
4. 5.664
5. 22.4
6. 55.2
7. 328.09
8. 10.024
9. 29.93
10. 1.242
Problem: 18.9 miles

Page 55

1. 217.2
2. 4.32
3. 1 1/8
4. 1 1/16

S. 2.52
S. 7.776
1. 11.52
2. .4598
3. .21186
4. .874
5. .1421
6. 9.7904
7. .0024
8. 1.904
9. 149.225
10. .0738
Problem: 11.25 tons

Page 56

1. 3 2/3
2. 1 1/2
3. 1 1/3
4. 8

S. 32
S. 7,390
1. 93.6
2. 72,600
3. 160
4. 736.2
5. 7,280
6. 70
7. 37.6
8. 390
9. 73.3
10. 76.3
Problem: $9,500.00

Page 57

1. 7 1/2
2. 7
3. 14/17
4. 10

S. 2.394
S. 15.228
1. 3.22
2. .464
3. .48508
4. 26
5. 3.156
6. 2,630
7. .0108
8. 3.575
9. 55.11
10. .04221
Problem: $41.37

Page 58

1. 19
2. 202
3. 1,001
4. 111

S. .44
S. 1.8
1. 19.7
2. 3.21
3. .57
4. 3.7
5. 6.08
6. 40.9
7. .324
8. .16
9. 2.12
10. 2.04
Problem: 21 pieces

Page 59

1. 2.18
2. .3
3. 18.83
4. 4.833

S. .0027
S. .019
1. .0007
2. .012
3. .056
4. .036
5. .043
6. .063
7. .023
8. .022
9. .003
10. .068
Problem: $5.51

Page 60

1. 34.2
2. .0932
3. 3
4. 3

S. .34
S. .06
1. .065
2. .62
3. 2.05
4. .15
5. .04
6. .04
7. .12
8. 1.575
9. .06
10. .418
Problem: 28

Page 61

1. .075
2. .026
3. 930
4. 9,300

S. 3.9
S. .24
1. 8
2. 170
3. .42
4. 80
5. 5.4
6. 3.2
7. 3.7
8. 3.2
9. 57.5
10. 9.2
Problem: 55.25 mph

Page 62

1. .8
2. 50
3. .38
4. 4

S. .5
S. .625
1. .6
2. .25
3. .4
4. .875
5. .55
6. .52
7. .625
8. .2
9. .2
10. .7
Problem: $15.00

Page 63

1. 10.96
2. 1.57
3. 2.184
4. .875

S. .005
S. 40
1. .76
2. .074
3. .025
4. 4.5
5. .24
6. 284
7. 33.1
8. 16.2
9. .4
10. .625
Problem: 117.25 pounds

Page 64 Review

1. 12.283
2. 10.36
3. 29.1
4. 20.6
5. 4.73
6. 4.57
7. 9.36
8. 54.4
9. .752
10. 1.39956
11. 236
12. 2,700
13. 1.34
14. 1.46
15. .65
16. 400
17. .05
18. .013
19. .875
20. .44

Page 65

1. 1 1/2
2. 3
3. 1 1/6
4. 7/15

S. 17%
S. 90%
1. 6%
2. 99%
3. 30%
4. 64%
5. 67%
6. 1%
7. 70%
8. 14%
9. 80%
10. 62%
Problem: 90 pounds

Page 66

1. 7%
2. 90%
3. 1 1/4
4. 7.75

S. 37%
S. 70%
1. 93%
2. 2%
3. 20%
4. 9%
5. 60%
6. 66%
7. 89%
8. 60%
9. 33%
10. 80%
Problem: 12.8 fluid ounces

Page 67

1. 3/4
2. 1 1/8
3. 3 7/10
4. 6 1/10

S. .2, 1/5
S. .09, 9/100
1. .16, 4/25
2. .06, 3/50
3. .75, 3/4
4. .4, 2/5
5. .01, 1/100
6. .45, 9/20
7. .12, 3/25
8. .05, 1/20
9. .5, 1/2
10. .13, 13/100
Problem: 3/4

Page 68

1. 1.8
2. .8
3. 1.872
4. 9.62

S. 17.5
S. 150
1. 4.32
2. 51
3. 15
4. 112.5
5. 32
6. 80
7. 10
8. 216
9. 112.5
10. 13.2
Problem: 1 1/4 gallons

Page 69

1. 11.05
2. .8
3. .03
4. 12

S. 3
S. 30
1. $56
2. $1,800
3. 9
4. $750
5. 16
6. $16,000
7. $4,200
8. $3.50/$53.50
9. 138
10. $1,650
Problem: 208.75 miles

Page 70

1. 88 r4
2. 900
3. 3,918
4. 96

S. 20%
S. 80%
1. 60%
2. 50%
3. 10%
4. 75%
5. 75%
6. 60%
7. 25%
8. 80%
9. 75%
10. 20%
Problem: 64 people

Answer Key

Page 71

1. 12.57
2. 8.857
3. 44.814
4. 4.7

S. 25%
S. 75%
1. 25%
2. 80%
3. 50%
4. 90%
5. 60%
6. 75%
7. 75%
8. 75%
9. 80%
10. 95%

Problem: $235

Page 72

1. 26.4
2. 60%
3. 27
4. 34

S. 75%
S. 40%
1. 80%
2. 75%
3. 25%
4. 80%
5. 95%
6. 20%
7. 48%
8. 20%
9. 98%
10. 75%

Problem: 24 correct

Page 73

1. 7%
2. 90%
3. 30%
4. .24, 6/25

S. 9
S. 25%
1. 3.6
2. 57.6
3. 75%
4. 80%
5. 77.5
6. 6
7. 25%
8. 10%
9. 450
10. 51.92

Problem: 80%

Page 74

1. 1/6
2. 5 3/4
3. 3 4/15
4. 2 2/5

S. 32
S. 25%
1. 300
2. 60%
3. $14.40
4. 80%
5. 75%
6. 14.4
7. 240
8. 10%
9. $1,200
10. 20%, 80%

Problem: 84

Page 75 Review

1. 17%
2. 3%
3. 70%
4. 19%
5. 60%
6. .09, 9/100
7. .14, 7/50
8. .8, 4/5
9. 12.8
10. 138
11. 72
12. 60%
13. 80%
14. 25%
15. 20%
16. 25%
17. $80
18. 180
19. 45%
20. 75%

Page 76

1. 8.22
2. 13.14
3. .0492
4. .6

S. answers vary
S. answers vary
1. answers vary
2. answers vary
3. F, D, E points
4. $\overleftrightarrow{AC}$, $\overleftrightarrow{BC}$
5. $\overleftrightarrow{FE}$, $\overleftrightarrow{FD}$
6. $\overrightarrow{AB}$
7. $\overline{FD}$, $\overline{ED}$, $\overline{FE}$
8. answers vary
9. answers vary
10. point E

Problem: 25 hours

Page 77

1. 1 2/5
2. 1/2
3. 1 1/2
4. 4

S. answers vary
S. answers vary
1. answers vary
2. answers vary
3. answers vary
4. answers vary
5. answers vary
6. answers vary
7. answers vary
8. answers vary
9. answers vary
10. $\angle$IHJ

Problem: 5 pieces

Page 78

1. 75%
2. 33.75
3. 45
4. 70%

S. answers vary
S. answers vary
1. answers vary
2. answers vary
3. acute
4. obtuse
5. right
6. straight
7. answers vary
8. answers vary
9. answers vary
10. answers vary

Problem: .34 meters

Page 79

1. 1 2/7
2. 23/3
3. 3/5
4. 8 3/4

S. acute, 20°
S. obtuse, 110°
1. right, 90°
2. obtuse, 160°
3. acute, 20°
4. acute, 70°
5. obtuse, 130°
6. acute, 50°
7. straight, 180°
8. obtuse, 160°
9. right, 90°
10. obtuse, 130°

Problem: 208 students

Page 80

1. 75%
2. 90%
3. 2.6
4. 40%

S. acute
S. acute
1. acute
2. acute
3. obtuse
4. obtuse
5. acute
6. acute
7. right
8. obtuse
9. obtuse
10. acute

Problem: 266 seats

Page 81

1. DEF, acute
2. FGH, obtuse
3. JKL, right
4. parallel

S. rectangle/parallelogram
S. triangle
1. square; rectangle; parallelogram
2. rectangle; parallelogram
3. trapezoid
4. triangle
5. trapezoid
6. square; rectangle and parallelogram
7. parallelogram
8. rectangle; parallelogram
9. triangle
10. trapezoid

Problem: $30.00

Page 82

1. 233 r23
2. 1,728
3. 1,155
4. 6,812

S. scalene/right
S. equilateral/acute
1. scalene; obtuse
2. isosceles; acute
3. isosceles; right
4. scalene; acute
5. equilateral; acute
6. scalene; obtuse
7. scalene; right
8. isosceles; acute
9. equilateral; acute
10. isosceles; right

Problem: 3 buses

Page 83

1. scalene
2. right
3. isosceles/acute
4. 45

S. 34 ft.
S. 30 ft.
1. 47 ft.
2. 48 ft.
3. 33 ft.
4. 54 ft.
5. 70 ft.
6. 41 ft.
7. 225 mi.
8. 34 ft.
9. 86 ft.
10. 34 ft.

Problem: 190 ft.

Page 84

1. 46 ft.
2. 56 ft.
3. 20%
4. 5/12

S. diameter
S. answers vary
1. radius
2. chord
3. $\overline{CD}$; $\overline{DE}$; $\overline{DG}$; $\overline{DF}$
4. $\overline{AB}$; $\overline{GF}$; $\overline{CE}$
5. 8 feet
6. point P
7. $\overline{RY}$; $\overline{VT}$; $\overline{XS}$
8. 48 feet
9. $\overline{PX}$; $\overline{PS}$; $\overline{PZ}$
10. $\overline{XS}$
Problem: 16 miles

Page 85

1. 350
2. 12.81
3. 4.1
4. 24.04

S. 12.56 ft.
S. 31.4 ft.
1. 18.84 ft.
2. 25.12 ft.
3. 28.26 ft.
4. 87.92 ft. or 88 ft.
5. 37.68 ft.
6. 31.4 ft.
7. 12.56 ft.

Problem: 75.36 ft.

Page 86

1. 19.52
2. 1
3. 3
4. 227.5

S. 169 sq. ft.
S. 165 sq. ft.
1. 84 sq. ft.
2. 400 sq. ft.
3. 132 sq. ft.
4. 10.75 sq. ft.
5. 6 sq. ft.
6. 625 sq. ft.
7. 6 1/4 sq. ft.

Problem: 182 sq. ft.

Page 87

1. 224 sq. ft.
2. 256 sq. ft.
3. 25.12 ft.
4. 43.96 ft.

S. 78 sq. ft.
S. 77 sq. ft.
1. 54 sq. ft.
2. 176 sq. ft.
3. 17 1/2 sq. ft.
4. 91 sq. ft.
5. 84 sq. ft.
6. 117 sq ft.
7. 71 1/2 sq. ft.

Problem: 94.2 ft.

Page 88

1. 800 sq. ft.
2. 45 sq. ft.
3. 98 sq. ft.
4. 169 sq. ft.

S. 50.24 sq. ft.
S. 113.04 sq. ft.
1. 78.5 sq. ft.
2. 615.44 sq. ft. or 616 sq. ft.
3. 12.56 sq. ft.
4. 200.96 sq. ft.
5. 78.5 sq. ft.
6. 113.04 sq. ft.
7. 153.86 sq. ft. or 154 sq. ft.

Problem: 75.36 ft.

Page 89

1. 52 ft.
2. 168 sq. ft.
3. 18.84 ft.
4. 28.26 sq. ft.

S. P=38 ft.   A= 84 sq.ft.
S. P=23 ft.   A= 20 sq. ft.
1. P= 48 ft.   A= 144 sq. ft.
2. P= 44 ft.   A= 120 sq. ft.
3. P= 38 ft.   A= 72 sq. ft.
4. C= 18.84 ft. A= 28.26 sq. ft.
5. C= 43.96 ft. (44 ft.)   A= 153.86 sq. ft. (154 sq. ft.)
6. P= 24 ft.   A= 24 sq. ft.
7. P= 32 ft.   A= 64 sq. ft.

Problem: 216 sq. ft.

Page 90

1. 1/10
2. 1 1/6
3. 10 1/2
4. 5

S. rectangular prism
    6 faces; 12 edges; 8 vertices
S. square pyramid
    5 faces; 8 edges; 5 vertices
1. cone, 1, 1, 1
2. cube, 6, 12, 8
3. cylinder, 2, 2, 0
4. triangular pyramid, 4, 6, 4
5. triangular prism, 5, 9, 6
6. sphere
7. 1

Problem: 40% girls; 60% boys

Page 91 Review

1. answers vary
2. answers vary
3. answers vary
4. answers vary
5. answers vary
6. answers vary
7. answers vary
8. answers vary
9. sides, equilateral,
    angles, acute
10. sides, scalene,
    angles, right
11. 28 feet
12. 28.26 feet
13. 153.86 sq. ft. (154 sq. ft.)
14. 98 sq. ft.
15. 126 sq. ft.
16. 58 1/2 sq. ft.
17. 256 sq. ft.
18. square, pyramid, 5, 8, 5
19. 140 sq. ft.
20. 384 feet

Page 92

1. 78 sq. ft
2. 40.82 ft.
3. 75%
4. 12.75

S. 3
S. -21
1. 14
2. -18
3. -14
4. -31
5. 24
6. 37
7. -168
8. -13
9. -34
10. -9
Problem: 336 students

Page 93

1. 2
2. -2
3. -34
4. 28.26 sq. ft.

S. -4
S. -7
1. -2
2. -7
3. -8
4. 14
5. -2
6. 10
7. -22
8. -13
9. -25
10. -84
Problem: won 75%, lost 25%

Page 94

1. 84 ins.
2. 1
3. 9
4. -3

S. -14
S. -15
1. 12
2. -3
3. 9
4. -28
5. 46
6. 21
7. -10
8. -11
9. -103
10. 15
Problem: $9.03

Page 95

1. -15
2. 3
3. 34
4. -108

S. -40
S. 15
1. -53
2. 11
3. -15
4. 15
5. 12
6. 3
7. -33
8. 15
9. 101
10. -681
Problem: 13 packages

Page 96

1. 3 3/4
2. 4
3. 12
4. 8

S. 48
S. -126
1. 68
2. -64
3. 288
4. -368
5. -736
6. -133
7. 21
8. 380
9. -256
10. 192
Problem: -10°

Page 97

1. 5
2. 10
3. 27
4. 252

S. 42
S. 54
1. -24
2. -48
3. 120
4. 360
5. -6
6. -24
7. -32
8. 72
9. 72
10. 330
Problem: $22.50

Page 98

1. 101 r1
2. 60
3. 86.1
4. 14.81

S. -3
S. 6
1. -16
2. 48
3. 15
4. -26
5. 22
6. -24
7. -6
8. -34
9. 13
10. -19
Problem: -26°

Page 99

1. -9
2. 8
3. -90
4. -24

S. 1
S. -12
1. -2
2. 3
3. -4
4. -36
5. -3
6. -2
7. 1
8. -1
9. 3
10. 2
Problem: 498 points

Page 100 Review

1. -2
2. 2
3. -16
4. -1
5. -19
6. -2
7. 13
8. -12
9. -27
10. -1
11. -48
12. 76
13. 56
14. 48
15. -9
16. 42
17. 16
18. 3
19. -2
20. 20

Page 101

1. 153.86 sq. ft.
2. 9.75
3. 1 1/2
4. 1 3/4

S. April
S. 10
1. July
2. July
3. June
4. 7° -8°
5. April
6. March; April
7. May; July
8. 13°
9. 3° -4°
10. March, April, July
Problem: 37.68 yards

Page 102

1. 75%
2. 70%
3. 192 sq. ft.
4. 220 r3

S. Winston/Auberry
S. 1,050
1. 1,200
2. 300
3. 350
4. 3,300
5. 300
6. 450
7. 1,450
8. 300
9. Sun City
10. 1,250
Problem: 20 parts

Page 103

1. 39 sq. ft.
2. scalene/obtuse
3. 2,365
4. 43.38

S. 90
S. 10
1. Tests, 4, 6, 7
2. 20
3. approx. 91
4. 4
5. 80, 85
6. 90
7. 15
8. 15
9. 3
10. improved
Problem: $26.25

Page 104

1. 1 2/5
2. 3/4
3. 1 6/7
4. 1 2/3

S. July/September
S. 100
1. 700
2. 100
3. May
4. June
5. 100
6. July, Sept.
7. 1,100
8. 200
9. 100
10. April, May
Problem: 85

Page 105

1. -44
2. 15
3. -21
4. 30

S. 23%
S. 70%
1. 33%
2. $460
3. $200
4. $400
5. 68%
6. 32%
7. $24,000
8. 80%
9. car, clothing
10. other expenses
Problem: 1,225 miles

Page 106

1. 3/4
2. 75%
3. 75%
4. -17

S. 1/6
S. 1/4
1. 5
2. 10
3. 8
4. 9/24 = 3/8
5. 16/24 = 2/3
6. 90
7. 6:00 A.M.
8. 2:30 P.M.
9. 40
10. 4/24 = 1/6
Problem: 168 sq.ft.

Page 107

1. 6,332
2. 30,628
3. 932
4. 1,003

S. 6,000
S. 4,500
1. 1989
2. 13,500
3. 1989; 1991
4. 7,000
5. 11,000
6. $150,000
7. $600,000
8. 5,000
9. 19,500
10. 3,000
Problem: $131

Page 108

1. 64 ins.
2. 18.84 ft.
3. 390 sq. ins.
4. 44 sq. ft.

S. 1990
S. 10 hours
1. 45
2. approx. 5
3. 2,000
4. 1960
5. 1950, 1960
6. approx. 2
7. 8
8. approx. 12
9. approx. 5 hours
10. approx, 10 hours
Problem: $26.25

Page 109 Review

1. Grant Falls
2. 525-550 feet
3. 100 feet
4. Snake Falls
5. Morton Falls
6. 13%
7. 20%
8. $600
9. $300
10. 41%
11. 71° - 72°
12. 30°
13. June
14. August
15. approx. 10°
16. 60,000
17. 40,000
18. 200,000
19. 180,000 pounds
20. salmon, cod, snapper

Page 110

1. 1,211
2. 539
3. 11,584
4. 162 r3
S. 106
S. 2,485 miles
1. 6,976
2. $209
3. 1,718
4. 267 votes
5. 9,000 miles
6. 59 mph
7. 276 miles
8. 833 feet
9. 272 pieces
10. 5, 172 books

Page 111

1. 1 1/10
2. 5/8
3. 7 1/2
4. 5
S. 3 17/20 cups
S. 11 pieces
1. 8 1/4 pounds
2. $40
3. 30 miles
4. 29 1/4 minutes
5. 34 feet
6. 2 2/3 pounds
7. 1 1/2 pounds
8. 13 3/4 hours
9. 125 miles
10. 16 tires

Page 112

1. $13.69
2. $3.44
3. $23.52
4. $2.19
S. $1.44
S. 540 mph
1. $ .41
2. $429.70
3. 2.34 sq. miles
4. $1.14
5. 192 miles
6. 6 pounds
7. $10.55
8. 264.85 pounds
9. $.89
10. 8 gallons

Page 113

1. 36.8
2. 5
3. 153.86 sq. ft.
4. 144 sq. ft.
S. 4 pieces
S. 16 cans
1. 556 feet
2. $3.36
3. 89
4. 8.025 inches
5. $28
6. 6 1/4 feet
7. $36.36
8. 40 students
9. 24,197
10. 270 cows

Page 114

1. 36 r1
2. 234 r2
3. 63 r7
4. 210 r15
S. 87
S. 965 lbs.
1. $9,500
2. 81 hours
3. 11,088 bushels
4. 254 students
5. 7 buses
6. 24 classes
7. 1,950
8. 81,000 gallons
9. 72 seats
10. 2,020

Page 115

1. 37
2. 1 3/4
3. 4 1/12
4. 6
S. 4 yards
S. 28 dollars
1. 1 1/4 gallons
2. 18 girls
3. $98
4. 39 bracelets
5. 2,000 acres
6. 300 miles
7. $48
8. 200 sq. ft.
9. 17 bushels
10. 1 7/20 pounds

Page 116

1. 19.87
2. 2.205
3. .353
4. $2.10

S. $16.90
S. $8.06
1. $10,500
2. $7.14
3. $107.95
4. $7.14
5. $6.90
6. $4.05
7. $12.84
8. $17.11
9. $840
10. $2.62

Page 117

1. 14.4
2. 25%
3. 37.68 ft.
4. 80 ft.

S. 64 seats
S. 12.1
1. $8.95
2. 23 miles
3. 64 boys
4. $18
5. $459
6. 8 pieces; $24
7. $50
8. 154 3/4 pounds
9. $6.72
10. $10.32

Page 118

1. scalene
2. right
3. 117 ft.
4. 256 sq. ft.

S. $32.19
S. 80 ft./$560
1. 480 miles
2. 39 hours 20 minutes
3. 14 students
4. $240
5. 510 students
6. $150
7. $830
8. $1,000,000, 500 lots
9. 200 acres
10. $74.82

Page 119

Final Review

1. 10,647
2. 60 mph
3. 14 boys
4. 5 pieces
5. 2 7/12 hours
6. 144 miles
7. $.69
8. $3.46
9. 79
10. 2,180 miles
11. 4 1/3 yards
12. 18 girls
13. $4.36
14. $2.75
15. $336
16. $2.76
17. 10 payments
18. 1,200 acres
19. $485.10
20. 272 ft.; $850

Page 120

Final Review

1. 814
2. 2,178
3. 23,965
4. 9,314
5. 7,693
6. 279
7. 3,628
8. 1,414
9. 1,714
10. 5,284
11. 292
12. 28,544
13. 1,665
14. 15,732
15. 158,178
16. 131 r2
17. 344
18. 14 r8
19. 253 r24
20. 69 r1

Page 121

Final Review

1. 5/7
2. 1 1/4
3. 14/15
4. 5 7/8
5. 14 3/10
6. 3/4
7. 3 1/2
8. 2 6/7
9. 5 1/10
10. 4 11/12
11. 6/35
12. 3/44
13. 20
14. 2
15. 8 3/4
16. 2 1/2
17. 4 2/3
18. 1 1/3
19. 3 2/3
20. 4 1/2

Page 122
Final Review

1. 11.913
2. 14.52
3. 30.7
4. 18.6
5. 2.61
6. 70.32
7. 7.92
8. 189.8
9. 1.512
10. .53298
11. 36.5
12. 3,600
13. 2.32
14. 1.34
15. .31
16. 300
17. .03
18. 13
19. .6
20. .35

Page 123
Final Review

1. 13%
2. 3%
3. 70%
4. 19%
5. 60%
6. .08; 2/25
7. .18; 9/50
8. .8; 4/5
9. 2.22
10. 128
11. 64
12. 80%
13. 75%
14. 80%
15. 80%
16. 60%
17. 256 girls
18. 26 games
19. 75% strikes
20. 10% incorrect

Page 124
Final Review

1. answers vary
2. answers vary
3. answers vary
4. answers vary
5. answers vary
6. answers vary
7. answers vary
8. answers vary
9. sides, isosceles, angles, acute
10. sides, scalene angles, obtuse
11. 51 feet
12. 18.84 feet
13. 28.26 sq. ft.
14. 54 sq. ft.
15. 98 sq.ft.
16. 38 1/2 sq. ft.
17. 196 sq. ft.
18. rectangular prism, 12 edges, 6 faces, 8 vertices
19. 96 sq. ft.
20. 148 ft.

Page 125

Final Review

1. -3
2. 3
3. -13
4. -5
5. -24
6. -3
7. 11
8. -12
9. -27
10. -4
11. -32
12. 57
13. 24
14. 36
15. -4
16. 52
17. 17
18. 4
19. -8
20. -8

Page 126

Final Review

1. grade 4
2. 1,500 cans
3. 9,500 cans
4. 21,500 cans
5. grade 2
6. 20 students
7. 11 more C's
8. 100 students
9. 13
10. B's
11. 25 inches
12. 9 inches
13. January
14. 60 inches
15. December
16. 500 boxes
17. Troop 15
18. 200 boxes
19. 900 boxes
20. 750 boxes

Page 127

Final Review

1. 872 miles
2. 435 mph
3. 20 correct
4. 6 bags
5. $2 9/10 = $2.90
6. 150.5 miles
7. $1.56
8. $6.75
9. 98 pounds
10. $290.50
11. 4 1/6 miles
12. 20 correct
13. $6.90
14. $11.19
15. 232 sq. ft.
16. $3.04
17. $600
18. 1/6 of pie
19. $81.15
20. 112 ft.; $1,540

Glossary

A

absolute value   The distance of a number from 0 on the number line. The absolute value is always positive.

acute angle   An angle with a measure of less than 90 degrees.

adjacent   Next to.

algebraic expression   A mathematical expression that contains at least one variable.

angle   Any two rays that share an endpoint will form an angle.

associative properties   For any a, b, c:
addition: $(a + b) + c = a + (b + c)$
multiplication: $(ab)c = a(bc)$

B

base   The number being multiplied. In an expression such as 4^2, 4 is the base.

C

coefficient   A number that multiplies the variable. In the term 7x, 7 is the coefficient of x.

commutative properties   For any a, b:
addition: $a + b = b + a$
multiplication: $ab = ba$

complementary angles   Two angles that have measures whose sum is 90 degrees.

congruent   Two figures having exactly the same size and shape.

coordinate plane   The plane which contains the x- and y-axes. It is divided into 4 quadrants. Also called coordinate system and coordinate grid.

coordinates   An ordered pair of numbers that identify a point on a coordinate plane.

D

data   Information that is organized for analysis.

degree   A unit that is used in measuring angles.

denominator   The bottom number of a fraction that tells the number of equal parts into which a whole is divided.

disjoint sets   Sets that have no members in common. {1,2,3} and {4,5,6} are disjoint sets.

Glossary

distributive property   For real numbers a, b, and c: a(b + c) = ab + ac.

element of a set   Member of a set.

empty set   The set that has no members. Also called the null set and written Ø or { }.

equation   A mathematical sentence that contains an equal sign (=) and states that one expression is equal to another expression.

equivalent   Having the same value.

exponent   A number that indicates the number of times a given base is used as a factor. In the expression n^2, 2 is the exponent.

expression   Variables, numbers, and symbols that show a mathematical relationship.

extremes of a proportion   In the proportion $\frac{a}{b} = \frac{c}{d}$, a and d are the extremes.

factor   An integer that divides evenly into another.

finite   Something that is countable.

formula   A general mathematical statement or rule. Used often in algebra and geometry.

function   A set of ordered pairs that pairs each x-value with one and only one y-value.
(0,2), (-1,6), (4,-2), (-3,4) is a function.

graph   To show points named by numbers or ordered pairs on a number line or coordinate plane. Also, a drawing to show the relationship between sets of data.

greatest common factor   The largest common factor of two or more numbers. Also written GCF. The greatest common factor of 15 and 25 is 5.

grouping symbols   Symbols that indicate the order in which mathematical operations should take place. Examples include parentheses ( ), brackets [ ], braces { }, and fraction bars —— .

hypotenuse   The side opposite the right angle in a right triangle.

I

identity properties of addition and multiplication   For any real number a:
addition: $a + 0 = 0 + a = a$
multiplication: $1 \times a = a \times 1 = a$

inequality   A mathematical sentence that states one expression is greater than or less than another. Inequality symbols are read as follows: $<$ less than
$\leq$ less than or equal to
$>$ greater than
$\geq$ greater than or equal to

infinite   Having no boundaries or limits. Uncountable.

integers   Numbers in a set. ...-3, -2, -1, 0, 1, 2, 3...

intersection of sets   If A and B are sets, then A intersection B is the set whose members are included in both sets A and B, and is written $A \cap B$. If set A = {1,2,3,4} and set B = {1,3,5}, then $A \cap B$ = {1,3}

inverse properties of addition and multiplication   For any number a:
addition: $a + -a = 0$
multiplication: $a \times 1/a = 1$ $(a \neq 0)$

inverse operations   Operations that "undo" each other. Addition and subtraction are inverse operations, and multiplication and division are inverse operations.

L

least common multiple   The least common multiple of two or more whole numbers is the smallest whole number, other than zero, that they all divide into evenly. Also written as LCM. The least common multiple of 12 and 8 is 24.

linear equation   An equation whose graph is a straight line.

M

mean   In statistics, the sum of a set of numbers divided by the number of elements in the set. Sometimes referred to as average.

means of a proportion   In the proportion $\frac{a}{b} = \frac{c}{d}$, b and c are the means.

median   In statistics, the middle number of a set of numbers when the numbers are arranged in order of least to greatest. If there are two middle numbers, find their mean.

mode   In statistics, the number that appears most frequently. Sometimes there is no mode. There may also be more than one mode.

multiple   The product of a whole number and another whole number.

Glossary

N

natural numbers   Numbers in the set 1, 2, 3, 4,... Also called counting numbers.

negative numbers   Numbers that are less than zero.

null set   The set that has no members. Also called the empty set and written Ø or { }.

number line   A line that represents numbers as points.

numerator   The top part of a fraction.

O

obtuse angle   An angle whose measure is greater than 90° and less than 180°.

opposites   Numbers that are the same distance from zero, but are on opposite sides of zero on a number line. 4 and -4 are opposites.

order of operations   The order of steps to be used when simplifying expressions.
1. Evaluate within grouping symbols.
2. Eliminate all exponents.
3. Multiply and divide in order from left to right.
4. Add and subtract in order from left to right.

ordered pair   A pair of numbers (x,y) that represent a point on the coordinate plane. The first number is the x-coordinate and the second number is the y-coordinate.

origin   The point where the x-axis and the y-axis intersect in a coordinate plane. Written as (0,0).

outcome   One of the possible events in a probability situation.

P

parallel lines   Lines in a plane that do not intersect. They stay the same distance apart.

percent   Hundredths or per hundred. Written %.

perimeter   The distance around a figure.

perpendicular lines   Lines in the same plane that intersect at a right (90°) angle.

pi   The ratio of the circumference of a circle to its diameter. Written π. The approximate value for π is 3.14 as a decimal and $\frac{22}{7}$ as a fraction.

plane   A flat surface that extends infinitely in all directions.

point   An exact position in space. Points also represent numbers on a number line or coordinate plane.

positive number   Any number that is greater than 0.

power   An exponent.

prime number   A whole number greater than 1 whose only factors are 1 and itself.

probability   What chance, or how likely it is for an event to occur. It is the ratio of the ways a certain outcome can occur and the number of possible outcomes.

proportion   An equation that states that two ratios are equal. $\frac{4}{8} = \frac{2}{4}$ is a proportion.

Pythagorean theorem   In a right triangle, if c is the hypotenuse, and a and b are the other two legs, then $a^2 + b^2 = c^2$.

Q	

quadrant   One of the four regions into which the x-axis and y-axis divide a coordinate plane.

R	

range   The difference between the greatest number and the least number in a set of numbers.

ratio   A comparison of two numbers using division. Written a:b, a to b, and a/b.

reciprocals   Two numbers whose product is 1. $\frac{2}{3}$ and $\frac{3}{2}$ are reciprocals because $\frac{2}{3} \times \frac{3}{2} = 1$.

reduce   To express a fraction in its lowest terms.

relation   Any set of ordered pairs.

right angle   An angle that has a measure of 90°.

rise   The change in y going from one point to another on a coordinate plane. The vertical change.

run   The change in x going from one point to another on a coordinate plane. The horizontal change.

S	

scientific notation   A number written as the product of a numbers between 1 and 10 and a power of ten. In scientific notation, $7,000 = 7 \times 10^3$.

set   A well-defined collection of objects.

slope   Refers to the slant of a line. It is the ratio of rise to run.

Glossary

solution   A number that can be substituted for a variable to make an equation true.

square root   Written $\sqrt{}$. The $\sqrt{36} = 6$ because $6 \times 6 = 36$.

statistics   Involves data that is gathered about people or things and is used for analysis.

subset   If all the members of set A are members of set B, then set A is a subset of set B. Written $A \subset B$.
If set A = {1,2,3} and set B = {0,1,2,3,5,8}, set A is a subset of set B because all of the members
of a set A are also members of set B.

U
---

union of sets   If A and B are sets, the union of set A and set B is the set whose members are included
in set A, or set B, or both set A and set B. A union B is written $A \cup B$. If set = {1,2,3,4}
and set B = {1,3,5,7}, then $A \cup B$ = {1,2,3,4,5,7}.

universal set   The set which contains all the other sets which are under consideration.

V
---

variable   A letter that represents a number.

Venn diagram   A type of diagram that shows how certain sets are related.

vertex   The point at which two lines, line segments, or rays meet to form an angle.

W
---

whole number   Any number in the set 0, 1, 2, 3, 4...

X
---

x-axis   The horizontal axis on a coordinate plane.

x-coordinate   The first number in an ordered pair. Also called the abscissa.

Y
---

y-axis   The vertical axis on a coordinate plane.

y-coordinate   The second number in an ordered pair. Also called the ordinate.

Important Symbols

<	less than	π	pi
≤	less than or equal to	{ }	set
>	greater than	| |	absolute value
≥	greater than or equal to	$.\overline{n}$	repeating decimal symbol
=	equal to	1/a	the reciprocal of a number
≠	not equal to	%	percent
≅	congruent to	(x,y)	ordered pair
( )	parenthesis	⊥	perpendicular
[ ]	brackets	| |	parallel to
{ }	braces	∠	angle
...	and so on	∈	element of
• or ×	multiply	∉	not an element of
∞	infinity	∩	intersection
a^n	the nth power of a number	∪	union
√	square root	⊂	subset of
Ø, { }	the empty set or null set	⊄	not a subset of
∴	therefore	△	triangle
°	degree		

Multiplication Table

x	2	3	4	5	6	7	8	9	10	11	12
2	4	6	8	10	12	14	16	18	20	22	24
3	6	9	12	15	18	21	24	27	30	33	36
4	8	12	16	20	24	28	32	36	40	44	48
5	10	15	20	25	30	35	40	45	50	55	60
6	12	18	24	30	36	42	48	54	60	66	72
7	14	21	28	35	42	49	56	63	70	77	84
8	16	24	32	40	48	56	64	72	80	88	96
9	18	27	36	45	54	63	72	81	90	99	108
10	20	30	40	50	60	70	80	90	100	110	120
11	22	33	44	55	66	77	88	99	110	121	132
12	24	36	48	60	72	84	96	108	120	132	144

Commonly Used Prime Numbers

2	3	5	7	11	13	17	19	23	29
31	37	41	43	47	53	59	61	67	71
73	79	83	89	97	101	103	107	109	113
127	131	137	139	149	151	157	163	167	173
179	181	191	193	197	199	211	223	227	229
233	239	241	251	257	263	269	271	277	281
283	293	307	311	313	317	331	337	347	349
353	359	367	373	379	383	389	397	401	409
419	421	431	433	439	443	449	547	461	463
467	479	487	491	499	503	509	521	523	541
547	557	563	569	571	577	587	593	599	601
607	613	617	619	631	641	643	647	653	659
661	673	677	683	691	701	709	719	727	733
739	743	751	757	761	769	773	787	797	809
811	821	823	827	829	839	853	857	859	863
877	881	883	887	907	911	919	929	937	941
947	953	967	971	977	983	991	997	1009	1013

Squares and Square Roots

No.	Square	Square Root	No.	Square	Square Root	No.	Square	Square Root
1	1	1.000	51	2,601	7.141	101	10201	10.050
2	4	1.414	52	2,704	7.211	102	10,404	10.100
3	9	1.732	53	2,809	7.280	103	10,609	10.149
4	16	2.000	54	2,916	7.348	104	10,816	10.198
5	25	2.236	55	3,025	7.416	105	11,025	10.247
6	36	2.449	56	3,136	7.483	106	11,236	10.296
7	49	2.646	57	3,249	7.550	107	11,449	10.344
8	64	2.828	58	3,364	7.616	108	11,664	10.392
9	81	3.000	59	3,481	7.681	109	11,881	10.440
10	100	3.162	60	3,600	7.746	110	12,100	10.488
11	121	3.317	61	3,721	7.810	111	12,321	10.536
12	144	3.464	62	3,844	7.874	112	12,544	10.583
13	169	3.606	63	3,969	7.937	113	12,769	10.630
14	196	3.742	64	4,096	8.000	114	12,996	10.677
15	225	3.873	65	4,225	8.062	115	13,225	10.724
16	256	4.000	66	4,356	8.124	116	13,456	10.770
17	289	4.123	67	4,489	8.185	117	13,689	10.817
18	324	4.243	68	4,624	8.246	118	13,924	10.863
19	361	4.359	69	4,761	8.307	119	14,161	10.909
20	400	4.472	70	4,900	8.367	120	14,400	10.954
21	441	4.583	71	5,041	8.426	121	14,641	11.000
22	484	4.690	72	5,184	8.485	122	14,884	11.045
23	529	4.796	73	5,329	8.544	123	15,129	11.091
24	576	4.899	74	5,476	8.602	124	15,376	11.136
25	625	5.000	75	5,625	8.660	125	15,625	11.180
26	676	5.099	76	5,776	8.718	126	15,876	11.225
27	729	5.196	77	5,929	8.775	127	16,129	11.269
28	784	5.292	78	6,084	8.832	128	16,384	11.314
29	841	5.385	79	6,241	8.888	129	16,641	11.358
30	900	5.477	80	6,400	8.944	130	16,900	11.402
31	961	5.568	81	6,561	9.000	131	17,161	11.446
32	1,024	5.657	82	6,724	9.055	132	17,424	11.489
33	1,089	5.745	83	6,889	9.110	133	17,689	11.533
34	1,156	5.831	84	7,056	9.165	134	17,956	11.576
35	1,225	5.916	85	7,225	9.220	135	18,225	11.619
36	1,296	6.000	86	7,396	9.274	136	18,496	11.662
37	1,369	6.083	87	7,569	9.327	137	18,769	11.705
38	1,444	6.164	88	7,744	9.381	138	19,044	11.747
39	1,521	6.245	89	7,921	9.434	139	19,321	11.790
40	1,600	6.325	90	8,100	9.487	140	19,600	11.832
41	1,681	6.403	91	8,281	9.539	141	19,881	11.874
42	1,764	6.481	92	8,464	9.592	142	20,164	11.916
43	1,849	6.557	93	8,649	9.644	143	20,449	11.958
44	1,936	6.633	94	8,836	9.695	144	20,736	12.000
45	2,025	6.708	95	9,025	9.747	145	21,025	12.042
46	2,116	6.782	96	9,216	9.798	146	21,316	12.083
47	2,209	6.856	97	9,409	9.849	147	21,609	12.124
48	2,304	6.928	98	9,604	9.899	148	21,904	12.166
49	2,401	7.000	99	9,801	9.950	149	22,201	12.207
50	2,500	7.071	100	10,000	10.000	150	22,500	12.247

Fraction/Decimal Equivalents

Fraction	Decimal	Fraction	Decimal
$\frac{1}{2}$	0.5	$\frac{5}{10}$	0.5
$\frac{1}{3}$	0.3	$\frac{6}{10}$	0.6
$\frac{2}{3}$	0.6	$\frac{7}{10}$	0.7
$\frac{1}{4}$	0.25	$\frac{8}{10}$	0.8
$\frac{2}{4}$	0.5	$\frac{9}{10}$	0.9
$\frac{3}{4}$	0.75	$\frac{1}{16}$	0.0625
$\frac{1}{5}$	0.2	$\frac{2}{16}$	0.125
$\frac{2}{5}$	0.4	$\frac{3}{16}$	0.1875
$\frac{3}{5}$	0.6	$\frac{4}{16}$	0.25
$\frac{4}{5}$	0.8	$\frac{5}{16}$	0.3125
$\frac{1}{8}$	0.125	$\frac{6}{16}$	0.375
$\frac{2}{8}$	0.25	$\frac{7}{16}$	0.4375
$\frac{3}{8}$	0.375	$\frac{8}{16}$	0.5
$\frac{4}{8}$	0.5	$\frac{9}{16}$	0.5625
$\frac{5}{8}$	0.625	$\frac{10}{16}$	0.625
$\frac{6}{8}$	0.75	$\frac{11}{16}$	0.6875
$\frac{7}{8}$	0.875	$\frac{12}{16}$	0.75
$\frac{1}{10}$	0.1	$\frac{13}{16}$	0.8125
$\frac{2}{10}$	0.2	$\frac{14}{16}$	0.875
$\frac{3}{10}$	0.3	$\frac{15}{16}$	0.9375
$\frac{4}{10}$	0.4		